DATE			

BAKER & TAYLOR

When can you see fireballs and whom should you contact if you spot one? When is it best to hunt for comets and meteors and whereabouts? How do you gauge the size of the coma in the head of a comet and estimate its degree of condensation? Clear and easy to use, this guide shows you how to make successful and valuable observations and records of comets, asteroids, meteors, and the zodiacal light. For each topic the historical background and current scientific understanding support a wealth of observational techniques.

Comet observers are shown techniques for search and discovery. They can learn how to make visual estimates of brightness and size, and how to make photographic studies of cometary heads and tails. Asteroid hunters will find a 'life list' of quarry and guidelines on how to search for these objects and then how to photograph or electronically image them. Fruitful photographic and electronic methods for studying meteors and meteor showers are provided. Visual and photographic techniques show you how to examine the often elusive zodiacal light. The more adventurous are provided with advanced techniques on how to make successful astrometric, spectroscopic, and electronic observations. This is rounded off with an invaluable list of centers world-wide to contact with your details of unusual sightings.

Observing Comets, Asteroids, Meteors, and the Zodiacal Light

The Practical Astronomy Handbooks are a new concept in publishing for amateur and leisure astronomy. These books are for active amateurs who want to get the very best out of their telescopes and who want to make productive observations and new discoveries. The emphasis is strongly practical: what equipment is needed, how to use it, what to observe, and how to record observations in a way that will be useful to others. Each title in the series will be devoted either to the techniques used for a particular class of object, for example observing the Moon or variable stars, or to the application of a technique, for example the use of a new detector, to amateur astronomy in general. The series will build into an indispensable library of practical information for all active observers.

Titles available in this series

1. A Portfolio of Lunar Drawings
   by Harold Hill
2. Messier's Nebulae and Star Clusters
   by Kenneth Glyn Jones
3. Observing the Sun
   by Peter O. Taylor
4. The Observer's Guide to Astronomy (Volumes 1 & 2)
   edited by Patrick Martinez
5. Observing Comets, Asteroids, Meteors, and the Zodiacal Light
   by Stephen J. Edberg and David H. Levy
6. The Giant Planet Jupiter
   by John H. Rogers
7. High Resolution Astrophotography
   by Jean Dragesco

Observing Comets, Asteroids, Meteors, and the Zodiacal Light

STEPHEN J. EDBERG
Jet Propulsion Laboratory, Pasadena

and

DAVID H. LEVY
Lunar and Planetary Laboratory, University of Arizona

CAMBRIDGE
UNIVERSITY PRESS

Published by the Press Syndicate of the University of Cambridge
The Pitt Building, Trumpington Street, Cambridge CB2 1RP
40 West 20th Street, New York, NY 10011-4211, USA
10 Stamford Road, Oakleigh, Melbourne 3166, Australia

Printed in Great Britain at the University Press, Cambridge

A catalogue record for this book is available from the British Library

Library of Congress Cataloging-in-Publication Data

Edberg. Stephen J.
Observing comets, asteroids, meteors, and the zodiacal light / Stephen J.
Edberg and David H. Levy.
    p.   cm. – (Practical astronomy handbook series ; 5)
Includes index.
ISBN 0 521 42003 2
1. Comets–Technique.   2. Asteroids–Technique.   3. Meteors–
Technique.   4. Zodiacal light–Technique.   I. Levy, David H.,
1948–   .   II. Title.   III. Series. QB721.E35   1994
523.6–dc20 93–44922 CIP

ISBN 0 521 42003 2 hardback

Dedication

For Janet, who has had to share me with Astronomy, and Aaron, Shanna, and Jordan too, and for my parents Joe and Sophie, who guided, encouraged, and supported me.

With love from Steve

For my brother Richard, my sister Joyce and her husband Larry, and my brother Gerry and his wife Audrey, who know so well what Astronomy means to me.

With love from David

Contents

Authors' notes

'Debris of the solar system,' we thought, had a rather nice ring to it. A single volume with a discussion of the observing possibilities of four seemingly unrelated but complex areas – comets, asteroids, meteors, and the dust tepees familiarly called the zodiacal light – was, for its authors, a highly desirable project.

This book has its genesis in two ideas. The first was a pair of observing manuals we wrote for the Astronomical League and the Association of Lunar and Planetary Observers called *Observe: Comets* and *Observe: Meteors*. In the years since their publication we thought that although these areas are really very different, requiring diverse skills and procedures, comets and meteors as objects in space are so related that a single book about them would be an interesting challenge. Adding asteroids and the zodiacal light to the project followed naturally.

The second idea, that we do this book together, was not a challenge at all. Since we met each other ten years ago, our mutual passion for the subject has kept our friendship brisk and our planning for the book lively and entertaining.

For two people who don't live that far apart we always joke about the distances we have to travel to visit each other. One precious memory is of sitting together at a conference in Heidelberg while Uwe Keller of the Max Planck Institute of Aeronomy displayed the very first clear image of the nucleus of Halley's Comet. As we looked transfixed at what the Giotto spacecraft had photographed, we saw for the first time that this comet was a place, a world, with depressions and a mountain on its land. But during another meeting, this time in Canada, we became so engrossed in a back-of-the-hall conversation about the latest comet that someone asked us to leave the lecture we were supposed to have been paying attention to!

We hope that the following pages will help to clarify a difficult and elusive subject. We have included numerous references throughout the text so that you, whether you are a beginner or an expert, will have further opportunities to expand your knowledge and practice. We have deliberately tried not to put our different writing styles into a melting pot, and you will no doubt find that the writing for much of the visual observing sections, which are Levy's forte, has a different style from the thoughts expressed in the sections on photography, spectroscopy, and the analysis of data, areas which are Edberg's strength.

And although we have made every effort to clean up any errors, we are aware that this is a subject with several points of view, and that others will disagree with some of what we say. For example, many good meteor observers feel that observing in groups is far less desirable than observing individually. We know this is a point of contention. Nevertheless, we have devoted a considerable amount of space in the 'Meteors' chapter to describing how to observe meteors effectively as observing teams. Our main justification is that observing that way is fun, and fun is ultimately what amateur astronomy is all about.

Now as to the errors: if you find one, each author blames the other for it.

Stephen Edberg
David Levy

Acknowledgements

To Steve Iverson, for his guidance of an enthusiastic amateur astronomer through courses in general science and chemistry, and his sponsorship and leadership of the astronomy club through Edberg's high school years;

To Bill Curtis and Tom Baur who made Edberg's summer at the High Altitude Observatory an education and fond memory; to Kitt Peak and Mount Wilson and Sacramento Peak solar observatories for all the practical experience they gave Edberg; and to Don Landman who inspired Edberg to greater inquiry high atop Haleakala;

To Steve Larson, Clark Chapman, Donald Davis, Stuart Weidenschilling, and Rick Greenberg, for their teaching and inspiration during Levy's years of association with them;

To Eugene and Carolyn Shoemaker, with whom Levy has been observing since 1989, and who have been exemplary teachers;

To astronomers Larry Lebofsky, Ray Newburn, Peter Millman, Ian Halliday, Bruce MacIntosh, and David Meisel, to meteor observers Norman McLeod III, Michael Morrow, Karl and Wanda Simmons, Pete Manley, and special thanks to Sophie Edberg, Murray Geller, and Rollin Van Zandt;

To Christopher Spratt, who provided assistance with the section on history and naming of asteroids, and to Brian Marsden, Charles Morris, Clifford Cunningham, Daniel W. E. Green, and Mark Coco for their suggestions;

To Simon Mitton and Cambridge University Press; as usual they have been efficient, professional, patient, and fun to work with;

Finally, we will always be indebted to Ted Bowell, first-rate asteroid scientist and friend, who named two of the many asteroids he has discovered 3672 Stevedberg and 3673 Levy.

1

Introduction

Consider the orbits of two solar system members – the Earth and a tiny rock the size of a grain of sand as it hurries from the blackness of space as far away from the Sun as Neptune, moving inexorably closer to the Sun. A million years ago this speck belonged to some comet out in the Oort Cloud, and somehow, perhaps by a gravitational perturbation by a passing star, this comet left the cloud and headed towards the Sun. At some point in its history, the comet approached a large planet, probably several times, and slowly evolved a new orbit that had it confronting the Sun every few years. A spectacular celestial scimitar was seen for a few weeks by observers on Earth, before it shrank and faded away.

At one of these returns the sand-grain-sized particle blew off its parent comet, becoming part of the comet's dust tail, and continued to orbit in the comet's trail as a meteoroid long after the tail faded away. Through many returns to the Sun, the particle orbited; each time its motion was affected by gravitational and electromagnetic forces. And then the tiny speck met the atmosphere of Earth, vaporizing away in a fiery plunge which attracted the wide eyes of a person, young at heart, who had never before seen the sky. In attracting the young pair of eyes, the meteor, in its final dash, revealed a heaven and created a new amateur astronomer.

The story may have other scenarios, with similar endings. Orbiting the Sun between Mars and Jupiter, a mountain-sized planet encounters another. The collision produces a mighty explosion and fragments, from small-mountain-size to dust grains, are scattered through space, each following its own orbit but all of them eventually returning, at various times, to their explosive point of origin. Gravitational perturbations act on the fragments and eventually a boulder-sized piece begins crossing the orbit of Earth. As the years go by the crossing occurs closer and closer to when the Earth is at the same point until late one afternoon a flash in the sky as bright as the Sun announces the arrival of a newcomer to Earth. A few lucky viewers are treated to anomalous sounds – crackling and popping – at the same time as the brilliant fireball is visible, while others follow the silent course of the fireball until it fades out and minutes later hear the rumbling of the falling meteorite as its sonic boom makes its way across the landscape. After numerous interviews with witnesses, searchers head off to find the meteorite and analyze its fragments, to learn more about its collision-of-origin and the origin of the solar system as a whole.

Some collisions with Earth are not so benign. A comet or asteroid only a few kilometers (miles) across can impact the Earth with disastrous results: it is believed that the era of the dinosaurs, the Cretaceous Period, ended with their extinction due to the impact of a comet or asteroid that left a crater off the present coast of Yucatan (Mexico), preceded by an impact near Manson, Iowa. We have but to look at the Moon, which suffers far less weathering and geologic activity than Earth, to see what our planet might look like due to collisions with solar system debris through the ages.

Even the smallest particles have their effects. Too light to move rapidly through Earth's atmosphere and burn up, they instead slow down and waft gently to the surface, perhaps acting as seeds for raindrops as they drop through the atmosphere. Their brethren, in uncounted numbers around the Sun, are visible as faint glows in the sky after sunset and before sunrise, providing astute observers with yet another phenomenon to study.

Truly, the phenomena associated with solar system debris could keep an observer occupied for a lifetime, without the need to venture out farther in space to survey the wonders of the galaxy and the rest of the universe. We hope you will enjoy, as much as we do, making the observations of comets, asteroids, meteors, and the zodiacal light that we describe in this book.

2

General observation techniques

Getting the most from observing solar system debris requires training and practice. Whether the observations will be made by use a retina or some other detector, they will be enhanced by finding the best observing conditions and making efforts to use all the techniques discussed below. After all, even photographers and photometrists have to aim their telescopes.

2.1 Dark adaptation, averted vision, and eye sensitivity

Many amateur astronomers don't realize the difference full dark adaptation makes in viewing the sky and faint objects. While twenty to thirty minutes are necessary for initial dark adaptation, significant increases in the eyes' sensitivity occur with extended stays in the dark. One to two additional hours produce a noticeable increase in the detectability of faint objects when observed from dark-sky sites. The long periods required for full dark adaptation do not mean you must sit around in a closet doing nothing. It does mean that trips into illuminated areas must be abandoned, and the use of bright red flashlights should be curtailed. Use muted red lights for note taking or map reading.

Low-light sensitivity is enhanced by avoiding strong Sun and fluorescent light (Miller, 1980 and Chou, 1992). A good pair of sunglasses serves to cut the strength of light outdoors. Military surplus or fluoroscopic red goggles serve very well in starting the dark adaptation process (even in daylight) and in maintaining adaptation after dark when visits to lighted areas are unavoidable. Use a dark hood over your head to screen extraneous sky light or light pollution during observations. This removes distracting peripheral light.

Averted or indirect vision is useful when trying to see an object at the limit of the eyes' sensitivity. Because the high resolution, color-sensitive cones of the retina are situated on the optical axis of the eye, it is possible to see fainter sources by looking 10 to 20 degrees away from the source, while holding attention on the source. This allows light to fall on the much more light-sensitive (but color-insensitive) rods which are found in greater concentration off-axis.

Averted vision should not be used for visual magnitude and coma diameter estimates of a comet because it is difficult to repeat positioning of the comet and comparison stars on the same area of the retina. Data for other projects, especially in making drawings, may be improved by using averted vision.

2.2 Atmospheric transparency and sky brightness

Many observations are sensitive to the transparency of the Earth's atmosphere and to background sky brightness due to artificial and natural sources. Observations of a comet's tail, the size and magnitude of its coma, the visibility of asteroids, meteors, and the zodiacal light, photographic exposures, and photoelectric measurements are among those that are strongly influenced by transparency. Observers will be well rewarded for using observing sites with good transparency and minimal sky brightness. Of course, little can be done when the explosive eruption of a volcano, such as El Chichon in the 1980s and Mount Pinatubo in the 1990s injects a layer of dust into the stratosphere, increasing the absorption of light by a factor of two.

2.3 Seeing and optical quality

'Seeing' is the term used by astronomers to describe the steadiness of the atmosphere. When the stars twinkle – scintillate – rapidly, it is a sign that the seeing is poor and images will be unstable. Fine detail, including stars and asteroids appearing as pinpoints, lunar and planetary detail, and jets and fountains in cometary comas will all be smeared during poor seeing conditions. When the seeing is good, not only is fine detail more easily seen but fainter objects may be seen since their light is more concentrated and, therefore, more distinguishable against the background sky. Finding and noting small, faint comets will be much easier in good seeing.

Finding good seeing depends not only on geography and weather but also on the telescope and the observer. Internal air currents can ruin images. These are generated if the telescope and the surrounding atmosphere are at different temperatures. Permit your telescope to cool to the ambient temperature for the best images. This can take an hour. Even when this is done, images may be poor if nearby sources of heat affect the air in front of the telescope aperture, and those sources can be the observer and companions: don't have warm bodies under the light path and don't warm up the telescope tube by holding it for long while aiming at a target.

Little can be done about the weather, except to track it long enough to become familiar with the patterns, including the position of the jet stream and the status of atmospheric temperature inversions, that promote or degrade the seeing quality. Observing sites should be located, when possible, in the middle of large areas where air temperature will be uniform. An island or peninsula in a lake or the center of a large meadow are examples of potentially good (if, perhaps, a bit impractical) observing sites. A line of sight over parking lots, building roofs, or rock formations that store the day's heat will likely yield very poor images.

For the same reason observers desire good seeing, good optical quality and alignment are important. Check defocused star images to confirm that they are circular and appear the same both inside and outside of focus. Try a variety of eyepiece brands and focal lengths at the observing session, making direct comparisons of star image quality. Select the ones yielding the best images. It was Stephen Edberg's experience with Comet Halley in April 1987 (14 months after perihelion) that a good eyepiece made it possible to distinguish (indeed, to find) the comet against the background stars, whereas with a poor eyepiece the comet and all the faint stars were equally fuzzy.

2.4 Scientific observing

Astronomical observing should be fun. It isn't necessary for amateur observers to contribute to science, but many find this increases their joy in observing.

To be scientifically useful, data must be delivered with a sufficient amount of background information to allow physical interpretation. Insufficient calibration data render the observations useless, or nearly so. It cannot be emphasized enough that acquiring data for scientific use that lack necessary calibration is a waste of valuable time and energy.

The techniques of astronomical investigation, whether they rely on the eye, camera, or electronics, must first be learned and then practiced regularly to maintain proficiency. Also, an ongoing series of synoptic observations (that is, observations to provide a continuing general view) is more valuable than scattered individual observations. It is best to do one project well, continuing one type of effort, rather than attempt a variety of projects. A variety of projects may be done well (though usually none is done as well as an individual one which receives all of an observer's attention), but the group is often less valuable than concentration in one area.

2.5 Record Keeping and Submission

Much of the fun and all of the science you can do as an astronomer requires that you keep written records of your observations. We strongly recommend that you maintain a bound notebook, available from stationery stores, to record permanently your thoughts and impressions, weather and sky conditions, and the details – scientific and emotional – of your observations, as we discuss in detail later. Such a record will make it easy for you to go back and compare observations made in the past with more recent ones.

Data taken on comets can be submitted to any of several organizations or publications, depending on the type of data obtained. Accurate measurements of a comet's position with respect to the background stars (known as 'astrometry') should be reported to the Minor Planet Center and/or the Central Bureau for Astronomical Telegrams. Information regarding the brightness of comets should be sent to the *International Comet Quarterly*. Photographs and drawings should be sent to magazines such as *Sky and Telescope* or *Astronomy* (or the *ICQ*), or to groups such as the Comets Section of the Association of Lunar and Planetary Observers (ALPO), or the Comet Section of the British Astronomical Association. (A list of addresses is in Appendix V.)

2.6 Finding objects in the sky

Seeing solar system debris in the sky can require luck, knowledge, or preparation, or some combination of those. In the case of meteors, you only need to be looking up in the right direction to see them. In the case of fireballs, you may not even need to be looking up because their brilliance often attracts your attention to them.

The zodiacal light pyramids appear above the sunset and sunrise horizons, and the related, faint counterglow of light called the gegenschein may be found, if conditions are right, opposite the Sun in the sky.

A quick scan of the sky will reveal a bright comet above the horizon, but more effort is necessary for average comets and virtually all the asteroids.

Finding the locations of the brightest asteroids, Vesta, Pallas, Ceres, and Juno, is easy. All you need to do is seek out their positions, either listed in an ephemeris or mapped in *Sky & Telescope*, *Astronomy*, or other periodical, the annual *Observer's Handbook* of the Royal Astronomical Society of Canada, the *Astronomical Almanac*, or in some other annual listing of positions. Brighter comets will often be featured by periodicals as well, and comprehensively in the *International Comet Quarterly*'s annual *Comet Handbook*. For fainter comets and asteroids, ephemeris predictions can be computed from orbital elements available from almanacs and other catalogs; some commercial computer programs compute ephemerides and may even plot positions for you.

The process of moving from figures on the printed pages of some almanac to an object in the sky is simple and elegant. For any hour or night an object's position is given in two sets of numbers that define its place in the sky in right ascension, the celestial equivalent of earthly longitude, and declination, which corresponds to latitude. Plot the object's position on a star atlas that shows stars of seventh magnitude or fainter. (Ideally, you should plot the object's position on a chart showing stars at least as faint, or better, fainter than the object's predicted brightness. This could mean using a photographic star atlas, such as one of those by Vehrenberg [1963 and 1971] or even the Palomar Observatory Sky Survey.) Since positions are often given for 0:00 UT (i.e., midnight Greenwich Mean Time) which may be mid-afternoon for you, you will either have to 'guesstimate' or calculate where the object will be when it is dark and the object is above the horizon at your location.

There are two approaches to calculating the position, i.e., interpolating the motion of the object. One is to plot the two positions of the object, before and after you want to see it. Then use a ruler to draw a line between the positions and measure the separation. Divide the line into equal segments of convenient length and divide that number of segments into the amount of time separating the two positions. This gives you a good idea of how fast the object is moving and where it will be through the course of the observing period.

The other approach is the mathematical equivalent of drawing on the atlas. Find the differences in right ascension and declination for the two times surrounding the time of interest. Now compute the rates of change in position on a regular scale, such as hourly (that is, if the ephemeris provides positions every three days, divide the positional differences by 72 [72 = 3 × 24 hours]). Keeping in mind your time zone difference from the Greenwich meridian, multiply the desired number of hours of motion by the rates of change and add those to the starting position, so you now know where the object is at the time desired. Plot this position on your atlas for use at the telescope.

The two methods described use linear interpolation, which will be adequate for most situations. More precise interpolation methods are described in the *Astronomical Almanac*, its *Explanatory Supplement*, and in other books on astronomical computing.

Now set up and aim your telescope. Setting circles will get you to the right part of the sky quickly, but the technique of star hopping described below will likely be necessary, if only on a smaller scale.

Begin with the nearest bright star and move from star to star in the right general direction, identifying each star with its position relative to other stars in their patterns in the atlas. When you get to what you think is the right field of view the process diverges somewhat for asteroids and comets. If you are looking for an asteroid, draw a map of the stars in the field, concentrating especially on the fainter stars right near what you think is the asteroid [Fig. 2.1]. Sketch in lines between or extending from pairs of stars so that any 'star's' motion will be apparent after some time has passed. A few hours later, or the following night, check the field again. If the object you have found is the asteroid, it should have moved, and the alignments you sketched earlier will have changed. Depending on where the asteroid is relative to the Earth and Sun, it might be moving fast enough that you can confirm motion within an hour or two. Be sure to indicate east–west, at least, on your sketches (due to Earth's rotation, stars appear to move from east to west in the field of view) so you can orient it correctly for comparison. Rotating a prism diagonal or simply re-aiming an altazimuth-mounted telescope will change orientations with respect to gravity.

If the object is a comet, its fuzzy appearance as well as its motion will aid

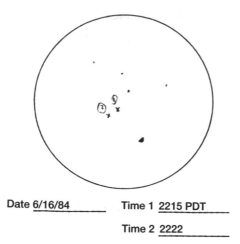

Date 6/16/84   Time 1 2215 PDT

Time 2 2222

Figure 2.1   Fast-moving asteroid 1984 KD was sketched on 17 June 1984. The asteroid was at Position 1 at 0515 UT and at Position 2 seven minutes later. Observation by S. Edberg.

in identification. Scan the field carefully looking for a patch of light. Use a higher magnification if necessary, sweeping the higher power field of view over the full area of the lower power field. Be careful! It is easy to mistake galaxies or small, unresolved groups of faint stars for comets, especially if you are anxious to find them. A comet filter *may* help you identify the comet, or at least increase its contrast against the background sky.

Two factors contribute to how fast an asteroid or comet appears to be moving: its motion and the Earth's motion. As an analogy, when you are driving down a highway, you will likely have occasion to pass a car moving slower. As you overtake the other car, it appears to slow down relative to you, and at the moment of passing it actually seems to move backwards. The asteroids and planets farther from the Sun than Earth (and hence moving slower than us) do the same thing. From night to night they appear to move east. Then, as the Earth 'overtakes' them, they slow down, almost stop, and appear to move backwards for several weeks until they slow down again and then resume their eastward motion.

You may be fortunate enough to observe an asteroid when it is moving rapidly east (direct motion) or west (retrograde). If it is at opposition from the Sun, a time when it rises and sets opposite the Sun in the sky and is above the horizon essentially all night, then its motion will be fairly rapid westward. If it sets early in the evening or rises in the morning sky, it will be moving eastwards. However, if it is within about a month or two of opposition from the Sun, then it will be 'stationary' and barely creeping across the background of stars. During this time its motion will be hard to determine. Confirming your sighting of an asteroid is a difficult process, in fact, if you have only allotted yourself one hour to do it. Asteroid observing needs time and patience, and you have to be certain of motion before you can add it to your list of asteroids observed.

2.7 Time

An important part of making astronomical observations is knowledge of the correct time. In some situations it is critical, and knowledge of it to the nearest fraction of a second is necessary, such as when a faint asteroid will be eclipsing a bright star. In other cases, the correct time to the nearest hour will suffice, for example when a slow-moving object is to be located.

The correct time is not as easy to come by as you might think. Telephone time signals are usually very close to correct, and some network radio broadcasts contain a synchronization tone exactly on the hour. To be absolutely certain of the time use a shortwave radio to pick up the broadcast radio time services of a government. In the United States the National Institute of Standards and Technology broadcasts over station WWV, transmitting at 2.5, 5, 10,

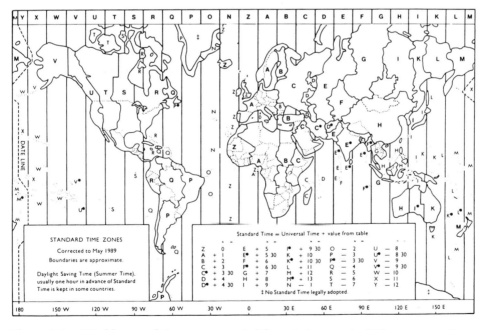

Figure 2.2 World map of time zones, copied from Astronomical Phenomena for the Year 1993. *Find your location on the map and use the letter to determine the difference between your time zone and Universal Time. Courtesy of the US Naval Observatory. SERC copyright. Reproduced with the permission of the Science and Engineering Research Council.*

15, and 20 MHz and over WWVH at 2.5, 5, 10, and 15 MHz. In addition, WWV time signals may be heard over the telephone at 303–499–7111 (Boulder, Colorado) or 900–410–8463 (50 cents/minute). Canada broadcasts time signals on CHU at 3.330, 7.335, and 14.670 MHz on the shortwave bands and over the telephone at 613–745–1576 (English language) and 613–745–9426 (French language). Japan broadcasts time signals on JJY at 2.5, 5, 8, 10, and 15 MHz. Australian broadcasts are on VNG at 2.5 5, 8.638 and 12.984 MHz continuously, and at 16 MHz from 22:00 to 10:00 UTC. Other nations broadcast time signals as well, including Germany (DCF77), the United Kingdom (MSF), South Africa (ZSR), and Russia (RWM). Stephen Edberg has occasionally picked up Japanese and Australian broadcasts in Lockwood Valley, California with his inexpensive Radio Shack Timecube® receiver.

Knowing the correct time is important, but knowing its relation to your local time is also important so you can determine if an event will occur during your night time. Figure 2.2 shows standard time zones and the relation of local standard time to Universal Time. Don't forget that some localities adopt summer or daylight saving time during a portion of the year.

3

Comets

3.1 For whom the comet tolls

This chapter is for anyone interested in observing comets. Both novice observers and advanced practitioners will find projects requiring observing equipment that ranges from unaided eyes to telescopes of any aperture. It is our hope that you will enjoy participating in this fascinating field at whatever level you choose.

Comet observing requires the full range of equipment mentioned above, all depending on the nature, size, and brightness of the comet you wish to observe, and what types of observations you wish to make. The only requirement for all observers is patience and a good telescope. For comet hunting, a wide field of view allows you to search large areas of the sky in a reasonable amount of time.

Figure 3.1 Comet Halley excited the public and was a pretty sight during the spring of 1986. Photograph by S. Edberg.

Even in a science where every field is known for unexpected surprises, comets are unusual. Their brightnesses and sizes, even their orbits are sources of the unexpected. The person who studies them is part of a lengthy line of astrologers, soothsayers, poets, natural philosophers, and, of course, many careful scientists. The line includes the woebegone scientist who dared defy his director at the Greenwich Observatory who forbade him to measure the Great Comet of 1861.[1]

It includes Jean-Louis Pons, whose record has at least 27 discoveries, and G. P. Bond, who observed from Harvard College Observatory as Donati's Comet completed its magnificent show. The line includes an early Native American who, a hundred centuries ago, must have gazed in wonder at some long-departed comet from the summit of Kitt Peak, and a terrified prince who prepared for his death at the sight of the bright thirteenth century comet.

The serious amateur comet observer is a person of commitment, driven in the past perhaps by terror, and now by wonder, curiosity, and a sense of scientific inquiry, at a strange apparition in the sky. Comets come and go, but the wonder, the unanswered questions, and the drive for another look stay with us. Comets are a part of us.

When a bright comet makes the national news, people wonder just as they did millennia ago. And when people gaze at a comet, they cannot be indifferent to this startling visitation by an object from far away. In times like these, people wish to learn more – how the comet was found, why it is here, and from where it came.

At such a time the comet observer, either with a small telescope and note pad or with the largest of telescopes and a computer, becomes a special commodity. And after the comet goes, the observer stays with telescope and star charts, and with luck, several faint and not-so-famous comets to study. These faint comets are, in a sense, the observer's lifeblood, for they are so much greater in number than the infrequent bright ones. They offer the individuality of special behavior, some with tails, a few even with anti-tails, some with visible nuclear condensations, and others appearing as featureless fuzzy patches. For the observer who specializes in the study of faint comets, the appearance of a spectacular comet is a special bonus.

3.2 Historical notes

The idea that a comet could be considered a warning from angry deities probably stems from actual appearances of comets during the great wars and sieges of our past. The theory of Aristotle, that comets were 'exhalations' in our own

---

[1] George Airy, director of Greenwich Observatory, was such a stickler for planning and detail that he was upset when observer William Ellis broke the observing plan and turned his telescope toward the unexpected Great Comet of 1861.

atmosphere, helped feed this fear. The last year of the siege of Troy was at one time believed to have been punctuated by an appearance of a comet, although the evidence is somewhat contradictory. A major comet appeared as far back as 371 BC, looking like 'a burning torch of extraordinary size', a fiery beam, lasting many nights.

A comet which appeared just one year after Caesar's death was somehow connected with the journey of his soul to the stars. For some political leaders, comets were not omens but licenses to terror. The comet of AD 64, according to Tacitus, afforded Nero the opportunity for more mayhem than was usual even for him. In 1181 a comet appeared to presage the death of Pope Alexander III, and only 17 years later another announced the passing of Richard I of England. In 1223, Philip Augustus of France passed away to the warning of a comet, and Pope Urban IV died with the appearance of a comet in 1254.

The appearance of Halley's Comet in April 1066, the time of the Norman Conquest, is, of course, the best known example of the importance that ancient and medieval peoples attached to comets. *'Nova stella, novus rex!'* ('New star, new king!') was the battle cry that highlighted the incredible influence that comets were believed to have had on world events. But there is at least one documented case of a comet's actually causing a person's death. This story comes from an August 1861 edition of *The Illustrated London News*:

> There is no doubt, however, that comets sometimes really did produce fatal effects. In June, 1402, one appeared in Italy which literally killed the famous John Galeas Visconti. The astrologers of the Prince had predicted that his death would be announced by a comet of extraordinary magnitude, and the celestial phenomenon had no sooner become visible than his Highness, speechless from fright, sank to the ground and died.

During the Renaissance, our slowly growing understanding of the nature of comets changed. Although the idea that comets might come from outside the atmosphere began to take hold thanks to the work of Tycho Brahe, it was Edmond Halley who proved the extraterrestrial nature of comets beyond any doubt by his brilliant work on the comet that now bears his name.

Halley's eighteenth century study was remarkable not just for what it revealed, but for his rigorous use of logical process in arriving at his conclusion. Still, it takes a spark of genius to note that among all the comets that have appeared since records were kept, one seemed to appear at 76-year intervals.

But even though the pace of maturity of cometary knowledge quickened as we approached our century, the old superstitions still survived, and the comets' perceived habit of occurring just before momentous events probably helped to prolong that superstition.

Just before Napoleon's invasion of Russia, the comet of 1811 arrived.

> 'One evening,' reported the nun Antonina, 'as we were on our way to a commemorative service at the Church of the Decollation de Saint-Jean, I suddenly perceived on the other side of the church what appeared to be a

13

resplendent sheaf of flame. I uttered a cry and nearly let fall the lantern. The Lady Abbess came to me and said, "What art thou doing? What ails thee?" Then she stepped three paces forward, perceived the meteor likewise, and paused a long time to contemplate it. "Matouchka," I asked, "what star is that?" She replied, "It is not a star, it is a comet." I then asked again, "But what is a comet?" The mother then said, "They are signs in the heavens which God sends before misfortunes." Every night the comet blazed in the heavens, and we all asked ourselves, what misfortunes does it bring?' (Guillemin, 1877).

3.2.1 *The nineteenth century*

Living in the last century would have to be a comet observer's dream. It began with the comet of 1811, one of the largest comets ever to visit the Earth's vicinity in historical times. Then in 1826, Wilhelm von Biela discovered a comet that proved to be periodic, with earlier independent discoveries in 1772 and 1805. During its appearance in 1846, the comet was noticed to have split into two (probably between 1840 and 1844), and astronomers watched the two comets slowly move apart from each other. At the pair's next return in 1852, the two comets appeared, but they were never seen again. It was widely assumed that the breakup marked the death knell to these comets, an idea that was substantially proven when, in 1872, the Earth encountered a storm of meteors from the comet, appearing at a rate of many hundreds per hour.

The comet of 1843 was wondrous. On 19 March its mighty tail stretched from just east of Cetus all the way south of Orion to Sirius in Canis Major, and was believed to have been almost 250 000 000 km (150 000 000 miles) long. Fifteen years later Donati's Comet hung in the evening sky, with two long tails, one of gas and the other of dust. Similar to Comet West of 1976, Donati's evening appearance made it more easily visible to the general population.

Many people independently found the Great Comet of 1860, with its 20-degree long tail (Kronk, 1984). But only a year later the comet of 1861, discovered from New South Wales by J. Tebbutt as a faint object barely visible to the unaided eye, brightened rapidly as it headed directly north. By the end of June, when it first became visible over England, it was so bright that sharp observers could see it before sunset. Its head, being close to Capella, was circumpolar from England, and its tail did not fade out until it was part way through Hercules. Imagine such a comet visible in the summer sky through the short summer night.

In 1877 Coggia's Comet grew a long tail that on 17 July stretched for some 45 degrees. Five years later a great sungrazing comet, the Great September Comet of 1882, broke into several pieces after its encounter with the Sun. These were spoiled times, a chance for people to live their lives from an apparition of one comet to that of another. It was a time for people, once and for all, to forget fears of comets and to turn to their study. In 1877, Amedée

Guillemin published *The World of Comets*, the first popular treatise that took comets seriously. Certain stories contained in this fascinating book indicate that our understanding of comets in the late eighteenth century still needed improvement:

> A French philosopher with a clever idea, Lambert, suggested that since comets are more numerous than planets, some of them may be habitable. In *'An Essay Upon Comets'*, Andrew Oliver in 1772 suggested that the purpose of a comet's tail, generated as the comet approaches the heat of the Sun, is to keep the temperature of the comet itself cooler for its inhabitants, and that as the comet recedes from the Sun, the tail and coma enwrap it, thus keeping the inhabitants warmer.

3.2.2 Bright twentieth century comets

Although our century has not boasted as many bright comets as its predecessor, it has not been totally disappointing. Comet Morehouse in 1908, with its extremely active tail, was followed by two major apparitions in 1910 – the Great Southern Comet, known as 1910a, and Halley's Comet. Then came a drought. Except for Comet Peltier in 1936 and a bright comet that was found during the November 1948 total eclipse of the Sun and was then visible in the following mornings in the constellation of Hydra, there were no spectacular comets. But the lull was broken in 1956 when Comet Arend–Roland presented a fine performance, complete with bright anti-tail pointing towards the Sun. A few months later a second comet, Mrkos, appeared bright in the evening sky.

Comet Ikeya–Seki, 1965f, is a member of a special group of 'sungrazers', to which the September 1882 comet also belongs. In the weeks after its discovery, astronomers suspected that 1965f might pass beneath the surface of the Sun. (Numerous comets discovered by the Solwind and Solar Maximum Mission satellites in the 1970s and 1980s did in fact collide with the Sun.) Fortunately for many east-coast observers who were plagued with bad weather and the southern declination of the comet's path, this special visitor grazed the Sun a mere 500 000 km (300 000 miles) above the Sun's surface, surviving to present northern hemisphere observers with a magnificent post-perihelion spectacle. Five years later, Comet Bennett delighted morning-sky observers. Comet Kohoutek, whose December 1973 perihelion passage was well publicized but disappointing to the public, provided comet scientists with a wealth of information about a comet's first passage around the Sun. But just two years later, Comet West became much brighter than expected, breaking apart and presenting a fine morning view which unfortunately was missed by a public poorly informed of this fine opportunity. In 1983, Comet IRAS–Araki–Alcock raced across a moonless sky, coming to within 5.5 million km (3.3 million miles) of Earth. It crossed the northern sky so quickly that observers without telescopes could see it moving in a period of a few minutes; observers with small telescopes could actually watch it move across the sky.

Despite the poor Earth-comet geometry of Periodic Comet Halley's 1986 perihelion, the scientific community mounted an all-out effort to observe this most famous of comets. Five spacecraft – two from Japan, two from the Soviet Union, and one from a consortium of European countries – encountered the comet as it crossed the plane of the Earth's orbit in March 1986, a few weeks after the comet's perihelion passage. The European spacecraft, Giotto, made the closest encounter, revealing a dark comet nucleus marked with a depression and a mountain-like feature. Meanwhile, public star-parties around the world gave many thousands of people their first look at a comet; city after city reported traffic congestion and long lines as people enthusiastically waited for a glimpse of the renowned comet. NASA set up the International Halley Watch to encourage and collect observations. Amateur astronomers contributed some 15 000 observations to the IHW, including magnitude estimates, drawings, photographs, and spectra.

In the post-Halley period Comet Wilson 1986l, Periodic Comet Brorsen–Metcalf 1989o, Comet Okazaki–Levy–Rudenko 1989r, and Comet Austin 1989c_1 gave fine presentations for observers with binoculars and small telescopes, and Comet Bradfield 1987s showed a prominent anti-tail for a short time. In August of 1990 Comet Levy 1990c journeyed across the sky as a prominent visitor. No match for the really bright comets of the last century, this comet tried to make up by being visible throughout the August nights as it crossed from Pegasus to Aquila. Brightening to third magnitude, it was visible to observers worldwide.

As this book was in final preparation, Periodic Comet Swift–Tuttle was finally recovered after an absence of 130 years. It was a fine binocular object, barely visible to the unaided eye, and had a nice tail before it dived behind the Sun. There was a flurry of media excitement when some projections of its orbit suggested it had the potential to collide with Earth in 2126. While the comet will be observed astrometrically for the next few years to refine its orbit, there is now a consensus that a collision will not occur on its next return (but keep an eye out for a spectacular meteor shower!)

We close this section with a comet that doesn't qualify as bright in the conventional sense at the time of writing, but the fireworks it promises to deliver during the last half of July 1994 make it worthy of discussion.

On 25 March 1993 the team of Carolyn and Eugene Shoemaker and David Levy discovered an oddly-shaped image on films made with the 18-inch Schmidt Telescope at Palomar Observatory. This was designated Comet 1993e, now known as Periodic Comet Shoemaker–Levy 9. The confirmation observations made by James V. Scotti of the Spacewatch Survey at Kitt Peak National Observatory indicated multiple nuclei for the comet. Later images obtained by Jane Luu and David Jewitt with a larger telescope at Mauna Kea show as many as 17 individual fragments, and as many as 22 have been reported. As seen from Earth these fragments were extended in a 'bar' [Fig. 3.2] 51 arcsec in length on 30 March 1993 that had lengthened to about 70 arcsec at the end of June (J. Scotti, private communication). A dust trail of material [Fig. 3.3] associated with the comet spanned 17 arcmin at this time, and tails of

Figure 3.2   Twelve individual nuclei are visible in this CCD image of Comet P/ Shoemaker–Levy 9. The field of view is 1.3 arc minutes square. This five minute exposure was taken with the 2.3-m (90-inch) telescope of the University of Arizona at Kitt Peak by W. Wisniewski on 28 March 1993.

about 1 arcmin in length extended from some of the nuclei. (See International Astronomical Union *Circulars* 5725, 5726, 5730, 5732, 5744, 5745, 5782, 5800, 5801, 5807, and 5822 for additional details.)

The orbit determined for this fragmented comet had been continually

Figure 3.3   The faint, extended train of dust, the bright nuclear 'bar', and the combined tails of the nuclei are visible in this image of Comet P/Shoemaker–Levy 9 made by J. V. Scotti with the 0.91-m (36-inch) Spacewatch Telescope of the University of Arizona at Kitt Peak. The sharp edge on the southern side of the comet suggests that smaller particles are being blown away by the pressure of sunlight. The train extends more than 16 arc minutes across the frame.

refined, with the situation changing rapidly. The solutions showed that it would almost certainly collide with Jupiter in late July 1994. Such a collision is a very rare event. Running the ephemeris backwards shows a very close approach to Jupiter in July 1992, the implication being that it had broken up during that earlier encounter. As viewed from Earth the impact will occur on the planet's far side but the Galileo and Voyager spacecraft have viewing angles permitting observation of the events.

Images from the Hubble Space Telescope indicate a diameter upper limit of perhaps 5 km (3 miles) according to Harold Weaver and his colleagues. Fragment size upper limits of 1 km may be more realistic. Spectra reveal no emission lines; this is consistent with sunlight scattered by dust.

The expected impact velocity is 60 km/s at 37 degrees south latitude. A large fragment will dissipate one to ten million megatons of energy in the atmosphere of Jupiter according to Sekanina (1993).

'Hydro-code' studies by Zahnle, *et al.* (1993), show that larger fragments, with 1–10 km radius, may penetrate to 5–100 bars depth, below the clouds, before exploding. A fireball may be visible tens of seconds after the event. Surviving dust may be thrown into the atmosphere or above. The impact zone on the planet, near the morning terminator, should rotate into view from Earth about an hour after an event.

The effects on Jupiter could persist for months . . . or may not be noticeable at all. Orbit projections suggest that one end of the extended dust train can miss the planet and remain in orbit around it or go into solar orbit.

Estimates of the luminous efficiency range from 0.1 percent to 10 percent. Terminal bursts are likely as each fragment is destroyed and may be intense enough to be reflected by the Galilean satellites and seen as brightenings at Earth of 0.2 to 2 stellar magnitudes for fractions of a second. However, the mechanics of impacts of this size and by cometary bodies are simply unknown, and scaling from fireballs observed in Earth's atmosphere is quite uncertain. The series of impacts lasts as long as six days from 16–22 July.

This is surely one of the most anticipated events of this decade and anyone with a telescope has the opportunity to observe the effects on Jupiter of this rare event.

3.3 Cometary phenomena and nomenclature

In the centuries that comets have been observed, surprisingly little has been learned about the details of their origin, evolution, and the processes occurring in them. Most of our detailed knowledge of comets has been acquired in the past few decades, and even this is patchy. The wild image of a long, flaming tail and a bright, starlike head is dramatic but hardly accurate for most comets. It is infrequent that we have a comet of such magnitude, and rarely do we see one of truly gigantic proportions.

Some comets are believed to originate in the Oort Cloud, a huge collection of remnant cometary nuclei from the formation of the solar system. This cloud of Sun-orbiting comets extends from perhaps 20 000 to beyond 150 000 astronomical units (AU; 1 AU = 145 000 000 km = 93 000 000 miles, the average separation of the Earth and Sun). There are no perceptible boundaries to this spherical shell, and the trillion or more (10^{12} to 10^{13}) comets there are so widely scattered that using the word 'cloud' is almost misleading. Gravitational perturbations by passing stars disturb some comets' orbits enough to start them on the long journey to the inner solar system. Planetary perturbations during one of a comet's infrequent visits may kick the comet completely out of the gravitational grasp of the Sun or may trap it in a smaller orbit.

The Kuiper Belt is an inner, planar extension of the Oort Cloud believed to contain one hundred million to ten billion (10^8 to 10^{10}) nuclei. It is believed to lie in the same plane as the planets and extend outward from the outer planets. Computer simulations show it could be the source of most short-period comets. An object designated 1992 QB$_1$ (see Section 4.2.2), with an orbit ranging from perhaps 40 AU to 60 AU from the Sun over 296 years, may be the first identified member of the Kuiper Belt. It is approximately 200 km (120 miles) in diameter and its color suggests a carbonaceous surface that has been bombarded by cosmic rays.

Periodic comets are those with well-known orbital periods less than 200 years long. Short periods are classified as those less than 20 years and intermediate periods range from 20 years to 200 years. Long-period comets are those with periods greater than 200 years.

3.3.1 Anatomy: how are comets made?

A comet can be separated into three components: the nucleus, the coma, and the tail. Meteors and the zodiacal light are generally agreed to be related to comets, and inactive comets often appear as faint asteroids.

The nucleus is the source of all cometary phenomena. This tiny member of the solar system, almost never directly observed in visual wavelengths, generates some of the largest phenomena (comet tails) and some of the smallest objects (dust particles and free molecules and atoms) observed in the solar system [Fig. 3.4].

The nucleus is believed to resemble a snowball with dust mixed in. Results of spacecraft flybys and intense ground-based observations of Halley's Comet in 1986 offer the alternative description, at least with Comet Halley, of a snowy dirtball, with a much greater concentration of dust than previously thought. Recent interpretation by Sykes and Walker (1992) of dust trails that were discovered by the Infrared Astronomical Satellite (IRAS) associated with short period comets suggests that for comets in general, 'frozen mudball' might be a better description than Fred Whipple's canonical dirty snowball. (Dust trails

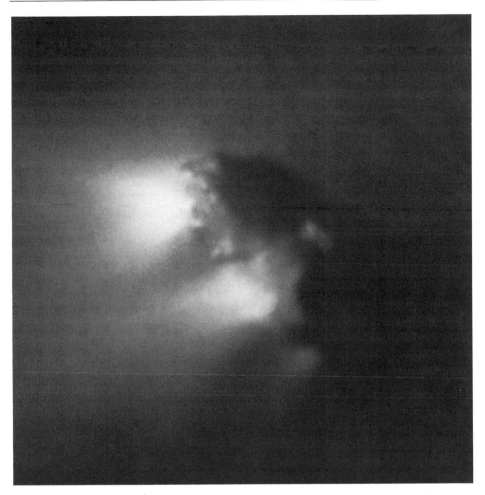

Figure 3.4   The nucleus of Halley's Comet as imaged by the Halley Multicolour Camera aboard the European Space Agency's Giotto spacecraft during its close flyby in March 1986. This is a composite image using the best resolution available for each part of the frame as the camera 'zoomed in' on the brightest source in its field of view. Note the jets and topographic features. Courtesy of H. U. Keller and © 1986 Max-Planck-Institut für Aeronomy, Lindau/Harz, Germany.

consist of relatively large particles that travel along the comet's orbit near the nucleus.)

The frozen volatiles in a nucleus, 'snow,' are not only pure water ice but, rather, a mixture of frozen gases including carbon dioxide (CO_2), hydrogen cyanide (HCN), and other gases containing carbon and sulfur. Some of these molecules are believed to be mixed with or trapped within the water ice and

21

dust. This picture has developed on the basis of spectroscopic studies of the molecules and the continuous spectrum of the tail.

It is hard to determine the proportions of gas and dust in a comet. The size of the nucleus is believed to range from a few hundred meters to 10 km or larger. The density of the nucleus is believed to be approximately that of water, one gram/cubic centimeter, but estimates range from 1/10 to two or three times that value. The occasional observations of the fragmentation of a cometary nucleus suggest that it has little internal strength.

During the comet's passage through the inner solar system, the heat of the Sun causes the ices to sublimate (change from solid to gas, directly, without changing to a liquid first). Comet Halley's nucleus loses material at a rate per orbital revolution that, if spread all over the nucleus, would be one meter (three feet) thick. In fact, portions of the nucleus are quiescent while active vents are the primary source of material in the coma. Nongravitational forces, a result of the 'rocket force' of the sublimation process, affect the comet's orbital motion to some extent.

The coma is an approximately spherical halo of material surrounding the nucleus. Gas streaming from the vents at 500 m/s (1650 feet/second) carries dust particles with it into this tenuous cometary atmosphere.

A coma does not generally form until the nucleus is within three astronomical units of the Sun. Faint comets usually generate smooth-appearing comas. More active comets like Halley often show jets or fountains emanating from the central condensation (the innermost, brightest portion of the coma) surrounding the invisible nucleus. Envelopes or hoods are often seen concentrically placed around the central condensation. Envelopes have been used to try to determine the direction of the rotation axis of Halley's Comet and its rotation period. This research is based on drawings of exceptional quality made during the 1835 apparition, on photographs from 1910 that have been digitally processed, and on CCD images from 1985 and 1986.

The coma and nucleus together are referred to as the head of the comet. Surrounding the head is a huge cloud of atomic hydrogen gas emitting ultraviolet light. Depending on the size and activity of the comet and its distance from the Sun, this hydrogen envelope can be one to ten million kilometers (six hundred thousand to six million miles) in diameter.

The Sun affects a comet in more ways than simply supplying heat to sublimate the nuclear ices. Electromagnetic radiation (radio, infrared, visible, ultraviolet, x-ray, etc.) from the Sun can interact with material released by the comet by affecting its electric charge and internal energy and by acting as a force to affect its motion after leaving the head. The solar wind and magnetic fields it carries play a role in shaping the tail.

A comet's tail is observed to have two components, both of which always start along the radius vector, the line from the Sun extending through the comet [Fig. 3.5]. The ion tail consists of molecules released by the nucleus that

Figure 3.5   Comet West exhibited well-separated gas (straight) and dust (broad, angled left) tails rising before the head during the early morning hours of March 1976. Note the poor corner images resulting from using the camera lens wide-open. Photograph by S. Edberg.

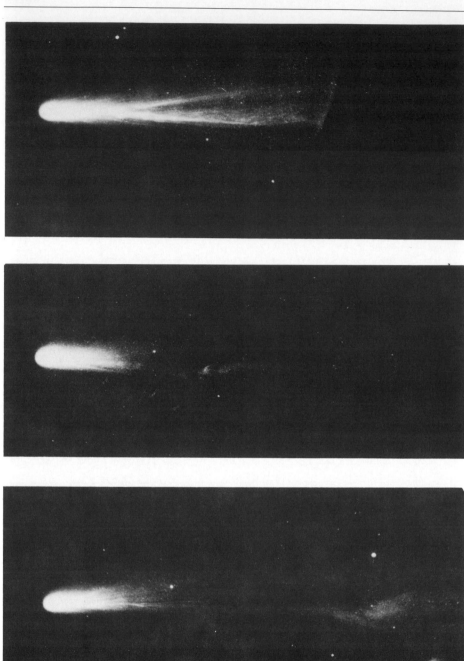

Figure 3.6   Comet Halley shows a disconnecting tail on 20, 21, and 22 March 1986. Photographs by R. P. Thicksten with the 18-inch Schmidt camera at Palomar Observatory. © 1986 California Institute of Technology.

have been ionized (forced to lose an electron and, thus, acquire a positive charge) by solar ultraviolet and x-radiation. It is bluish in color photographs and is usually quite straight, always extending away from the comet's head within a few degrees of the radius vector.

The solar wind, a stream of electrically charged particles (ions and electrons) blowing out from the Sun at several hundred kilometers per second, carries magnetic fields which drag cometary ions along with them away from the Sun. Solar electromagnetic radiation continuously excites the molecular ions causing them to glow in characteristic wavelengths. When light from the ion tail is broken down with a prism or diffraction grating, it shows an emission spectrum characteristic of a hot, thin gas with bright 'lines' primarily at the blue end of the spectrum.

On a bi-weekly or weekly basis, the ion tail of an active comet may appear to separate itself from the head [Fig. 3.6]. This 'disconnection event' appears to occur because the polarity of the magnetic field in the solar wind changes, and the weak cometary magnetic field reacts to this change. A new ion tail can be rebuilt in as little as 30 minutes after the disconnection. The disconnected old tail may remain visible (in photographs, especially) for several days after the event. Even when the tail doesn't disconnect, it may show changes over time scales of hours to days [Fig. 3.7].

Figure 3.7   This daily sequence of photographs of Comet Levy, taken over 20–28 August 1990, shows the constantly changing tail.

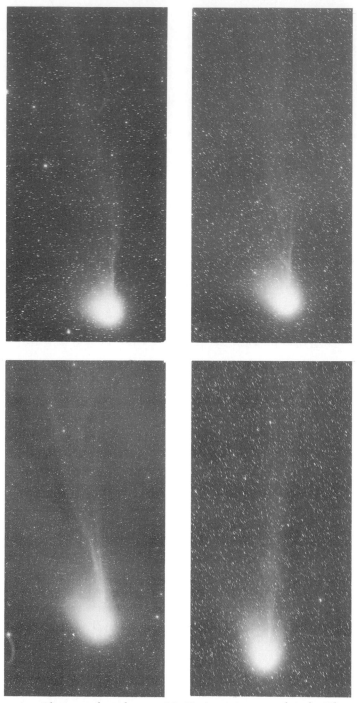

Figure 3.7 cont.   Photographs taken on 20–21 August were taken by Eleanor Helin. The remainder were taken by D. Levy with the 18-inch Schmidt camera at Palomar Observatory.

Figure 3.7 cont.

The dust tail is generated by a different process. Dust about one micrometer in size and, in part, silicate in composition is carried into the coma by the 'wind' of molecules released as the nucleus sublimates. Solar radiation pressure affects dust particles the way wind drives a sailboat. The dust is pushed away from the Sun while moving with the comet's orbital motion at the same time. The result is a curved dust tail. Dust is spread in the plane of the comet's orbit outside the orbital path.

When the comet is observed above or below the plane of its orbit, the dust and ion tails are well separated because ions respond strongly to the high-velocity, almost radial solar wind. For a few days around the time that the Earth crosses a comet's orbital plane, projection effects may present an anti-tail which appears to point in the direction of the Sun [Fig. 3.8]. The anti-tail is due to tail dust in the orbital plane well behind the comet; the dust appears opposite the tail because of the viewing angle.

Because sunlight illuminates the dust tail, a solar spectrum is scattered back to observers across its broad sweep. This tail may appear yellowish or even distinctly reddish in binoculars.

Banding or streaks photographed in the dust tail may be synchrones, groups

Figure 3.8   Comet Levy 1990c developed an anti-tail in February 1991. The anti-tail is narrower than the main tail, which is fan shaped and faintly extends (in position angle) all the way to the anti-tail. Photograph by E. and C. Shoemaker and D. Levy with the 18-inch Schmidt camera at Palomar Observatory.

of particles released at the same time, or striae, formed from parent particles released at the same time which later disintegrate and spread out in space. The leading edge of the tail is usually close to a syndyname, the locus of points along which all the particles respond to equal force.

Long after a periodic comet has left the inner solar system, its effects can still be observed. Particles released by the comet during its return to perihelion are affected by radiation pressure and the gravitational fields of the planets. Most micron-sized and smaller particles are blown out of the solar system. Larger particles will eventually fall into the Sun due to their interaction with sunlight, a demonstration of the Poynting–Robertson effect. Eventually, the largest particles, submillimeter size and larger, may be perturbed out of the comet's dust trail, closely matching its orbit, into orbits that intersect the Earth's. When the Earth crosses paths with these particles, 'falling stars' known as meteors are seen as the particles burn up in the Earth's atmosphere. Particles from the comet's tail slowly spread throughout the solar system and scatter sunlight (spread light in all directions, sometimes more strongly in preferred directions).

Multitudes of particles in the one-micrometer to 1/10-millimeter size range may be the source of the pyramidal glow on the sunrise or sunset horizons known as the zodiacal light. Recent studies suggest that their origin is from the collisions of asteroids. There are not enough comets observed to account for all the dust needed to make the zodiacal light.

3.3.2 Families of comets

Comets are members of families if they share, at some time in their history, a capture experience by a major planet. Jupiter has captured some fifty comets which have, on a single passage, approached the giant planet closely enough that its gravity altered the comet orbit. A comet on an orbit which earlier was parabolic would have its period altered drastically. Encke's Comet is probably the best known member of the Jovian family, and its period is some 3.3 years. In 1992 observations of Periodic Comet Wilson–Harrington showed that it is a Jupiter-family comet with a period of only 4.3 years, the shortest of any periodic comet except Encke (IAU *Circulars* 5585 and 5586). Although it was recorded in 1979–80 by Eleanor Helin as an asteroid, and again in 1988–89, it was not identified at those times as a comet; apparently this comet is usually inactive. (Some comets are often indistinguishable in appearance from asteroids, as the story of 1990 UL3 and Periodic Comet Shoemaker–Levy 2 (1990p) in Section 4.3.3 suggests.)

(a)

(b)

(c)

Figure 3.9.   Comet Halley showed considerable variation in its appearance with time. (a) On 18 November 1985, near the Pleiades, it showed no tail. A narrow gas tail developed during December and January. (b) By 21 March 1986, 1.5 months after perihelion, a fully developed gas and dust tail was enjoyed by observers. (c) The dust tail turned into a broad fan as seen in this photograph taken on 9 April 1986, as the Earth and comet made their closest approach and the view was along the axis of the tail. Photographs made by larger instruments recorded Halley's gas tail as well. By late April, through June, Halley displayed a long (though shortening with time), narrow tail. Photographs by S. Edberg.

3.3.3 Sungrazing comets

In a set of solar system objects noted for character, the sungrazers stand out as exceptional. Their extremely small perihelion distances, often less than 1.5 million kilometers (900 000 miles), distinguish this class of comets. The Kreutz group of sungrazers is notable for its origin in the break-up of one great comet possibly more than ten thousand years ago (Marsden, 1967).

Included in this group are the Great Comet of 1882, Comet Pereyra 1963e, Comet Ikeya–Seki of 1965, Comet White–Ortiz–Bolleli of 1970, and a number of comets found by the orbiting Solwind and Solar Maximum Mission satellites. Since these comets share a similar orbit, their apparent brightness and grandeur depends on two factors: the size of the comet and the time of year that it reaches perihelion. Pereyra, rounding the Sun in July 1963, was bright but far from spectacular. Both the 1882 and the 1965 comets rounded the Sun not far from our equinox times, and put on splendid shows. These two comets almost certainly split from a single body at their last perihelion, which may have been that of the Comet of 1106 (Marsden, 1967).

3.3.4 Comet nomenclature

A comet may be known by as many as three different designations. Immediately on its acceptance as a new or a recovered periodic comet by the Central Bureau for Astronomical Telegrams, the International Astronomical Union's clearinghouse for new astronomical discoveries, it is assigned a designation that includes the year followed by a lower case letter of the alphabet indicating the order of its acceptance at the Bureau for that year. In those years when the number of comets exceeds 26 (1987, 1989, and 1991, most recently), the alphabet is started again with a subscript $_1$ added for each designation. For instance, the ninth comet recognized in 1982 was designated 1982i and the twenty-seventh discovered or recovered in 1991 is $1991a_1$. Only periodic comets that are observed throughout their whole orbits do not receive this designation, including among others, P/Encke, P/Machholz, P/Gunn, P/Arend–Rigaux, P/Smirnova–Chernykh, and P/Schwassmann-Wachmann 1. Such comets are sometimes called annual comets.

A comet is named after its discoverer – up to three independent discoverers, or two discoverers working together. Prolific comet hunters like William Bradfield should have their comets referred to by name and designation to avoid confusion, like Comet Bradfield 1992b and Comet Bradfield 1992i. The discoverer need not be human as demonstrated by Periodic Comet Hartley–IRAS 1983v, which was first discovered by astronomer Malcolm Hartley and independently by IRAS, the Infrared Astronomical Satellite. A recovered, or newly discovered periodic comet (i.e., one that has a known period of orbital revolution less than 200 years) is known by the name or names of its discoverers, preceded by a P/. For example, it is correct to say Periodic Comet Halley, or P/Halley, but redundant to say both.

If a discoverer or discovery group finds more than one periodic comet, an arabic numeral follows the name. There is a Periodic Comet Levy since Levy has found one periodic comet by himself so far. But Periodic Comet Shoemaker–Levy does not exist. It became P/Shoemaker–Levy 1 when the second periodic comet was found, so now there is a Periodic Comet Shoemaker–Levy 1 and, in fact, eight others up to Periodic Comet Shoemaker–Levy 9. Periodic Comet Shoemaker–Levy 8 is the eighth discovered by the team of Carolyn and Eugene Shoemaker and David Levy. (Incidentally, all of the Shoemaker–Levy comets were discovered by Carolyn Shoemaker on films taken by the team of Eugene and Carolyn Shoemaker and David Levy. Thus it is possible to get your name on a comet if you are a part of the discovery team, even though you did not actually find it.)

Occasionally a periodic comet that had been considered lost (since it wasn't seen for several predicted apparitions) is accidentally rediscovered and then identified as the missing comet. It may then be assigned the rediscoverer's name in addition to the first discoverer's name, as happened recently with 1991a, P/Metcalf–Brewington. Periodic Comet Metcalf had not been seen since its discovery by J. H. Metcalf in 1906 until Howard Brewington, an amateur astronomer hunting from New Mexico, rediscovered it in 1991.

A handful of comets are named not for their discoverers but for their orbit computors. The most famous of these is P/Halley which has been well-observed since at least 240 BC. It was finally identified as a comet appearing every 76 years or so by Edmond Halley in 1705. Two others are P/Crommelin and P/Encke.

Some time after the end of each year all comets reaching perihelion in that year receive a Roman numeral designation giving the order in which they passed perihelion. P/Halley in 1910 was the second comet to reach perihelion that year and is known as 1910 II. P/Halley is also known as 1759 I and 1835 III, as well as 1982i, and now, following its 1986 perihelion, it is also referred to as 1986 III. Occasionally comets are discovered on old photographic plates and are assigned roman numeral designations following the better observed comets of that year, if an orbit can be determined. On rare occasions a comet is discovered by so many observers that no one receives priority as discoverer. Such a comet is usually very bright and is generally known by some other name. Examples include the Great Comet of 1843, and the Eclipse Comet of 1948.

3.4 Comet hunting

3.4.1 Background

A pivotal area in which amateurs can offer significant contributions to science is the search for and discovery of new comets. Historically, the amateur comet discoverer has been a source of pride and respect to his or her community, and stories of 'rags to riches' comet finders have graced the history books of two centuries.

Jean-Louis Pons began his observing career as a doorkeeper at an observatory in France. After some success as a discoverer of comets he was promoted to a full staff position, and his career total, conservatively estimated at 27 comet finds, has only recently been matched by Carolyn Shoemaker.

Leslie Peltier, an amateur astronomer from Delphos, Ohio, began his comet hunt with a borrowed 6-inch refractor. When he learned that his new instrument had an honored past, with three finds early in the twentieth century by Zaccheus Daniel, he decided to continue its tradition. Three years later, on 13 November, 1925, he discovered his first comet, and over the next three decades he found 11 more, one of which became easily visible to the naked eye. Peltier's (1965) autobiography, *Starlight Nights*, is a symphony to the joy of pursuit of comets, and reading it may give you a feel for the project you are about to undertake.

Kaoru Ikeya of Japan represents another success story. As a worker in a piano factory, he had to support his family whose name his father had disgraced. He found his first comet on 2 January 1963, and his second came some 18 months later. On 18 September 1965, Ikeya discovered an eighth magnitude patch of haze in Hydra, an object that Tsutomu Seki discovered independently only one hour later. Shortly after the discovery of Comet Ikeya–Seki was announced, an orbit was calculated that projected the object to approach within three hundred thousand miles of the Sun's surface, resulting in the most spectacular comet in more than half a century.

William A. Bradfield discovered his first comet in 1972, and his second in 1974, just after the passage around the Sun of the well-publicized Comet Kohoutek. Although it was more favorably placed for observation than Comet Kohoutek, the second Comet Bradfield, at least as far as the public was concerned, was lost in Kohoutek's publicity. In the next year Bradfield discovered two more, and two more again in 1976. He found his sixteenth, 1992i, in early 1992.

By discovering his first comet in the spring of 1978 with a large, 16-inch (40-cm) reflector, Rolf Meier became the first successful Canadian comet hunter. In 1979 he discovered a faint 12th magnitude comet, one of the faintest visual discoveries ever, until he exceeded it in 1984 when he found one even fainter.

Hunting for comets is a way of learning the sky. A comet search program is a concept designed for learning and enjoying the entire sky, star by star, galaxy by galaxy, on its own terms. A typical comet, if there be such a thing, lacks the bilateral symmetry of a spiral galaxy, the resolution of the star clusters. Discovering a comet has been a holy grail for amateurs since Messier began his program in the 1750s, and is a symbol of the amateur's ability to contribute to science. Amateurs have kept up their good record over many years despite the threat of professional takeovers, first with photography and wide field cameras and later with satellites. The hunt is as much sport as science, as observers display their skill in being the first to recognize the faint, perceptibly-moving patches of light.

The key to success in comet hunting is your approach to the program and the dedication with which you carry it out. You are undoubtedly familiar with the sightseeing process that occurs at star parties or in your back yard when you move your telescope from one object to another. Here you are in charge; you decide what you want to look at next. In an esoteric sense the sky is in charge in comet hunting, and every night is a surprise. You begin a night of observing with the joyful thought that you don't know just what the sky will offer you. Is tonight the night you'll get to see M57 or the Dumbbell Nebula, or is this to be a night of rich double stars and open clusters? By choosing the area of sky, you naturally have some influence over what you might see, but you are not exactly certain.

Why do you want to hunt for comets? Before you begin what may be a long-term, time-consuming project, you should ask yourself this question. There are several plausible reasons, the most obvious of which is that you want to discover a comet. We hope that is not your only reason. Comets are fascinating things to study and enjoy, and they are much more than mere targets of simple ego gratification.

Looking for a comet is like searching for a needle in the proverbial haystack. But in comet searching, the haystack is often more interesting than the needle which may be hiding within. When Don Machholz discovered his first comet, he had searched methodically for 1700 hours, an incredible amount of time with his eye at the eyepiece, the equivalent of searching through 140 twelve-hour nights, all under a clear moonless sky. This time does not include setting up or checking out suspicious objects, so if we add the total amount of time this project occupies in Machholz's life, the figure would be even more astounding.

Rolf Meier, an observer from Ottawa, Ontario, proposed an experiment to test the usefulness of searching with a large reflector of 16-inch (40-cm) aperture as opposed to traditional 6-inch (15-cm) or 8-inch (20-cm) telescopes. With four finds in less than 200 hours of searching, Meier proved his point.

Edward Emerson Barnard's nineteenth century program enjoyed a small bonus; at the time, an award of US $200 was made for each find, and Barnard was able to make his payments on a new house by discovering comets just as payments came due. So much of his home was financed through comet money that it came to be known as the 'comet house.'

It helps to be a bit of a poet. Once your telescope is set up, and your eye sees the first field of starlight that the telescope offers, you might feel a sense of deep relaxation as your mind turns from earthly thoughts to a sightseeing tour of the sky. What will the next field bring – a strange double star or a star that is unusually red, or blue? Will the next field show a strange planetary nebula, or globular cluster, or even a field of sprawling spiral galaxies?

If you spend your comet hunting time convinced that the next area must yield a comet, your hunt may be a disappointing activity indeed. The 'do-or-die' approach is not always compatible with a successful comet search, and unless you are a glutton for punishment we advise against it.

On the other hand, if you approach an evening's hunting with the attitude that you are about to spend an hour or two with your telescope searching for comets, and that you may find something interesting, you might have a much more enjoyable time. Usually the nights are successful that way. If your telescope is large enough (16 inches [40 cm] or more) you can identify many galaxies, clusters, and nebulae as non-cometary before you even have to go to the authority of a star atlas or catalog, so your hunt should be a pleasant and enriching experience. If your hour's work actually yields a comet, then the surprise and glory are yours. But don't begin the hunt expecting instant success. There is an interesting statistic that you must wait an average of 400 hours for your first find and an average of 200 hours for every subsequent one, and, over the centuries, that has somehow remained a reasonable average time. However, looking for 1000 hours or more is not unheard of for a first comet; your results will vary depending on you, your instrument, and your choice of observing location.

Luck also has something to do with success as a comet finder, although the more often you search, the luckier you should become. It is true that a seeker could spend a lifetime hunting comets without any success. Such failure could be due to bad luck, but a lack of good procedure or skill also has something to do with it. Sometimes a find truly comes by accident; the second and third discoverers of Comet Kobayashi–Berger–Milon were looking for M2 when they found an elusive fuzzy spot that turned out to be the new comet.

While bright comets tend to get discovered within 90 degrees of the Sun, a comet may appear in any part of the sky. One night David Levy was discussing this question with *Sky and Telescope* columnist Walter Scott Houston, out in the country at 2 a.m. under a crystal clear sky – only we were inside warming up. 'Isn't it true,' I asked innocently, 'that the best time to search for comets is during the last hour before dawn?' Houston's reply: 'The only thing that is really true is that you will never find a comet as long as we are inside talking about it.'

3.4.2 Preparation

Before you begin, it is helpful to familiarize yourself with the different types of deep sky objects like galaxies, clusters, and nebulae. The Messier Catalogue of nonstellar objects represents every type of deep sky object, offering excellent training for a would-be comet hunter. Finding these objects at different times of the year and under varying observing conditions will forewarn you to the discipline that is needed in comet hunting, a practice that does not come overnight, and that builds upon a special love of the sky that cannot appear from nowhere. If you can present the sky with evidence that you are serious (so to speak), then the sky in turn will reward you with a great variety of special things.

This is more of an invitation to study the Messiers than a warning to stay away from hunting. The Messiers are a *cordon bleu* of the finest deep sky objects one can behold; little of what is most delightful in the deep sky does not find a home in the Messier Catalogue, from M1, the fascinating Crab, to the magnificent Whirlpool galaxy, M51. The exquisite wisps of M42, the Orion Nebula, the nearest spiral galaxy, M31, the spectacular globular M13 in Hercules, the Ring Nebula M57; even the mysterious misshapen galaxy M82 – all these superlatives describe the grandeur of the Messiers. If you don't find a good proportion of them before you begin a serious comet search, then you are denying yourself the appropriate introduction to what your searching nights could offer.

While studying Messiers will give you an indication of what comets are not, observing comets that are already present and accounted for will give you an idea of what they are. Using positions and descriptions from such sources as the *Circulars* from the International Astronomical Union's Central Bureau for Astronomical Telegrams, keep an eye on the comets you can see with your instrument.

3.4.3 Patterns of search

Although comet hunting can be done with almost any telescope, instruments with short focal lengths are better for two reasons: first, for a given aperture they enable you to see more of the sky in one look, shortening the search time, and second, their wide fields increase the contrast between nebulous patches and sky background, increasing the chance that you will spot a faint comet. Conventional wisdom has always recommended altazimuth over equatorial mounts, since the simpler variety gives you more freedom to search comfortably down to the horizon. The equatorial mount, so useful in tracking objects as they cross the sky, becomes an encumbrance as you search for the easiest way to move up and down the sky in leisurely sweeps.

The search is far more effective if you follow an organized plan. Search a small area thoroughly, for example, rather than a large one haphazardly. Also, it is better to search for one hour for each of ten nights through areas more likely to have comets (the ecliptic at dawn or dusk) rather than spending ten hours on one night, which would force you through sections of the sky far from the Sun and unlikely to yield comets. Most important, never give up. Robert Burnham, who discovered several comets during the 1950s and 1960s, once said that 'if you hunt for comets long enough, sooner or later you will find one.

3.4.4 Equipment required

Your telescope needs to be capable of at least a three-quarter degree field of view at low power. Exactly how large the telescope should be depends on a number of factors. How much money do you wish to spend? Do you plan to mount the telescope at a dark site, or will you need to carry it around from place to place? The size of your telescope will also determine the brightness ranges of the comet you are likely to find. For comets of eighth magnitude or brighter, a 6-inch *f*/4 will suffice. But such a telescope would not catch anything much fainter, since the diffuse quality of some comets makes them more difficult to see than clusters and many nebulae of the same brightness.

3.4.5 Methods of hunting

There are as many variations on procedure for comet hunting as there are successful comet hunters. Rolf Meier hunts with an equatorially mounted telescope, checking suspicious objects quickly from their right ascension and declination. He never overlaps fields, and moves quickly at the rate of about a degree a second. William Bradfield (1981) makes horizontal sweeps with an altazimuth-mounted telescope over a wide arc of from 50 to 90 degrees, and always in the same direction. Other hunters use a zigzag pattern, either horizontally or vertically sweeping through the sky. On one point most hunters agree: patience, skill, and luck are necessary ingredients for a successful comet search.

Depending on the peculiarities of your own telescope and its mount, you might try different ways of hunting. You can hunt up and down strips of sky in a zigzag pattern, although this procedure is not recommended with an altazimuth telescope since, as the sky rises or sets, your search area will have gaps in it. You can sweep horizontally along a strip that is at some altitude above the horizon, or you can hunt along specific strips of declination or right ascension. What procedure you follow is not nearly as important as making certain that your search area is thoroughly covered. With his 16-inch (40-cm) reflector, David Levy hunts in an up-and-down pattern partly because his telescope moves more easily in altitude than in azimuth. He also hunts across certain blocks of sky in declining order of cometary importance. Thus, the comet-rich area within 90 degrees of the Sun in the evening sky is the first to get searched, as soon as the waning Moon has moved away from it. The other prime area, in the morning sky before dawn, is especially fertile. It is a place that has been hidden from our view by solar glare for months, and which may well harbor undiscovered comets. Not only the place, but also the time is right for hunting here. The adage that 'it is darkest just before dawn' makes practical sense; pollution from the previous day has had a chance to dissipate, some of the lights of evening have been extinguished, and the environment is much quieter than it was the previous evening.

The further away you get from the Sun, the less your chance to discover bright comets, and the greater your competition with the large photographic telescopes of professional astronomers.

3.4.6 When to hunt for comets

Comet hunting requires some judgment in choosing the best times for sweeping. On successive nights after full moon, you will see more and more hours of dark sky. Since the areas closest to the Sun have the best chance of containing an undiscovered comet, begin searching in the western sky in the nights following full Moon. Until moonrise you have a chance to look through a sky that has been blocked by moonlight long enough to mask the approach of a comet that is getting brighter as it approaches the Sun or Earth. The best time to hunt in the evening sky is just after the waning Moon has revealed a dark hour in the evening sky, an event that takes place some two days after full Moon. It is important to start your hunt when the western sky is dark enough to show at least fourth magnitude stars to the naked eye. There still may be light in the west, but your sky is getting better and so is your dark adaptation. Stop hunting when the Moon rises in the eastern sky.

The morning searches begin as the waning lunar crescent is thin enough not to interfere with the dark sky. The best time for comet sweeping is undoubtedly in the morning sky in the last two hours before dawn, beginning a few days before and continuing until about five days after new moon. Four times as many comets are discovered in this part of the sky as in the corresponding dusk regions. There is also a vaguely defined barrier, at about 100 degrees from the Sun, that separates most amateur from professional search areas.

The goal is to try to observe as much of the sky each month as possible, beginning with the regions most likely to produce comets and continuing into the less fertile areas. Overlapping and repeating searches during a month is a good policy, especially in the sky nearest the Sun, for you may easily have missed a faint comet. We recommend that you hunt a large area before repeating it.

Although amateurs have found comets near opposition from the Sun, their chances for such a find are low. These comets are the prey of the big photographic telescopes and are usually fainter than 12th magnitude.

3.4.7 Where to hunt

3.4.7.1 Sun vicinity

Comets can be discovered when the Sun is above the horizon. The procedure is simple and requires only a sharp pair of eyes and a clear, deep blue sky free of haze, cirrus clouds, and volcanic dust. You simply place an object, such as a street lamp, between you and the Sun, and scan the area within 20 degrees of our star. BE CAREFUL NOT TO EXPOSE YOUR EYES TO DAMAGING, DIRECT SUNLIGHT.

Comets have been discovered this way, most notably the bright comet 1910a, which had kept itself under wraps by approaching the Sun from the opposite direction from us. When it rounded the Sun it was discovered quickly by anyone who happened to look towards it. Such finds, it should be pointed out, are extremely rare and also obviously competitive, for a comet bright enough to be seen in daylight will undoubtedly be discovered independently by other observers. By searching in the Sun's vicinity, you might find such a bright comet first.

It is also important to ensure that your object is a comet, and not a cirrus cloud that is close to the Sun. Watch the size, shape, brightness, color, and apparent motion of anything you suspect might be a comet. Before reporting your observation, obtain at least two sightings, noting the time to the nearest minute UT, the distance in degrees from the Sun's center, and the clock angle of the object measured with the Sun at center and twelve o'clock as vertical. Try to get confirmation of your sighting from another observer.

3.4.7.2 Horizon at twilight

Searching for bright comets that are visible in the evening sky as it darkens, with either the unaided eye or with binoculars, can be rewarding. With the unaided eye, scan the horizon to some 30 degrees from the sunset point, about a half hour after the Sun has gone down. You may have a better chance for success by following the reverse procedure about a half hour before sunrise, since more comets tend to be found in the sky before sunrise than after sunset.

With binoculars, sweep the western horizon in the evening, or the eastern in the morning, from the horizon to an altitude of about fifteen degrees. Sweep slowly, with deliberate hand movements every two seconds from field to field. Although comets can appear at any time and any location, an evening search is best done just after the full Moon has left the area, so the two to five nights just after full Moon are the ones most likely to bring success. The nights around new Moon are the most promising for hunts in the dawn sky.

Any suspects should immediately be plotted on a sketch of the star field, noting the time, and then confirm that the object is not a background deep sky object plotted on a star atlas, or listed in a catalog, or the predicted return of a known comet.

3.4.7.3 Areas to avoid

If you use a small (under 25 cm (10 inches)) telescope, you might consider avoiding the cluster-rich areas of the Milky Way, and definitely overlook the galaxy areas of Coma Berenices, Leo and Virgo. If you dare to penetrate these 'sacred' areas, you'll be eyeball-deep in galaxies to check, some of which will look like faint comets.

The Milky Way contains so many stars that small concentrations of them might look like comets through a small telescope. However, if you hunt with a larger telescope, say 14 inches (35 cm) or more, you may not have to avoid the Milky Way, for the increased aperture you are using will allow you to identify most of its detailed structure. All the open clusters will of course be easily resolved, and the brighter globulars will unveil their secret identities too. You still have to check out some of the planetary nebulae and the faint globular clusters. In the galaxy-rich areas, you may be able to satisfy yourself that your fuzzy object is a spiral galaxy by its bilateral symmetry, although you may still have to check out the identities of most of the elliptical and irregular galaxies, and any galaxy with a low surface brightness, whose magnitude is not much more than that of the surrounding sky. David Levy has found that with his 16-inch (40 cm) the Coma–Virgo group is 'safe' to hunt through, and that he can identify by sight as non-cometary some 75 to 80 percent of the galaxies. But there are still many objects to check, and as a seeker of comets, you may find that region slow-going as you confirm each of the galaxies you spot. A few observers have memorized all the galaxies in these regions, and can use this knowledge to search for interlopers.

3.4.8 Ways to ruin an otherwise pleasant observing session

Wind After setting up your telescope, you find that the wind blows the tube around so much that you couldn't keep a comet, or anything else, in the field of view no matter how hard you tried. The better mounted your telescope is, the higher the wind speed must be before it puts you to bed; usually a wind of 20 miles per hour is as high as an observer can take comfortably.

Dew You breathe on the eyepiece, or condensation forms on one of the optical surfaces of your telescope, ending any chance of finding anything for a while. A hair dryer helps to dry the optics, but make sure that the air from the dryer is not hot enough to risk breaking the glass. Resistive heaters mounted on the telescope can provide a long-term solution to dew-fogging (Lucas, 1987 and Blackwell, 1992).

Technical Knockout You spot a fuzzy patch, then rush to your atlas to determine its identity. When you return to your telescope, the field of view, complete with its strange hazy spot, has moved off and you can't find it again. You continue searching, hoping that wasn't a new comet you just lost. You should always get a good idea of the object's position, using star patterns both in the telescope and the finder, or setting circle positions, before leaving the telescope to check an atlas. A 1x sight like the Telrad or Televue® Starbeam will help you get your bearings, and an adjustable 3 prong star pointer (Canady, 1991) will help transfer the sky position to an atlas. A quick photograph of the area, made with a co-aligned telephoto lens and ISO 3200 film will record the field permanently for more leisurely identification (S. Larson, private communication).

41

3.4.9 Eleven statistics regarding comet discoveries

Early in 1985 Don Machholz produced an interesting statistical study of the 33 comets that had been discovered by amateurs between the start of 1975 and the close of 1984. The work he produced provides some interesting statistics regarding amateur comet discoveries.

(1) These 33 comets represent a significant minority of the 162 comets that were discovered or recovered (found on the basis of a prediction) during the period.

(2) During the decade, an average of 3.3 comets were discovered each year by amateurs, of an average of 16.2 new and returning comets.

(3) Conducting a study of the brightnesses of the professional comet discoveries, Machholz determined that 'amateur astronomers would have found perhaps five more comets if the professional astronomers stopped discovering all comets'. These include Comet West, which, had it been missed, might have been picked up as a 10th magnitude glow at the end of 1975, Comet Hartley–IRAS (1983v), which was indeed discovered independently by David Levy, and Comet Shoemaker (1984s) which could have been found through amateur instruments shortly after its discovery.

(4) Comets do not get found at regular intervals. The shortest interval was essentially no time at all, as comets 1985j and 1985k were being found almost at the same time one morning; the longest involved two comet visual discovery 'droughts' lasting 18 months each.

(5) A large proportion of evening finds occurred in the period from 3–7 days past full Moon, with a second large proportion occurring just before new Moon. The morning discoveries were more evenly distributed from just before last quarter to three days before the next full Moon, with a peak around first quarter.

(6) The 'average' morning discovery took place 30.73 minutes before the beginning of astronomical twilight, while the 'average' evening discovery occurred 75.47 minutes after its end.

(7) The average comet found in the evening was magnitude 10.2, compared to 8.5 for morning finds. Machholz suggests that, among other factors, the morning sky is less intensely covered by searchers, thus allowing comets to brighten more before discovery.

(8) Although the heights above the horizon at which comets are found varied widely, the average for evening finds was 24.6 degrees, and for morning 28.3 degrees. Machholz makes the interesting point that most discoveries took place when the comet was above typical obstructions and horizon haze.

(9) Twenty-six amateurs, all male, discovered comets during the decade.

William Bradfield led the list with 10 finds, Rolf Meier followed with 4, Shigehisa Fujikawa had 3, and 5 observers had two each.

(10) Although the average hunting time per comet was 281.7 hours, the variation in times was enormous. They ranged from about an hour for Mori's second find to about 1700 hours for Machholz's first comet.

(11) Visual discoveries were made with reflectors of sizes 102 mm (4 inches) once, 145 mm (5.8 inches) once, 152 mm (6 inches) four times, 157 mm (6.2 inches) once, 202 mm (8 inches) three times, 254 mm (10 inches) twice, 404 mm (16 inches) five times, and 483 mm (19 inches) once. Discoveries were made with refractors of sizes 82 mm (3.3 inches) once, 127 mm (5 inches) once, and 152 mm (6 inches) eleven times. Only one discovery was made with a Schmidt–Cassegrainian telescope, of 202 mm (8 inches) aperture.

If you can begin a comet hunt in a serious way, you are in for at least five guaranteed treats, although finding a comet may not be one of them. (1) You will reacquaint yourself with a good proportion of the Messier objects. (2) You will be introduced to more objects of the *New General Catalogue* than you ever thought your telescope could pick up. (3) You will learn the distribution in the sky of types of deep sky objects, and note anomalies (like the rare winter globular clusters). (4) You will experience the telescopic meteor showers that often do not coincide with the naked eye ones. (5) Finally, you may possibly pick up some known comets, and make some useful observations that will teach you the subtle differences in light that tell a comet apart from its more distant kin.

3.4.10 Photographic search procedures

Photographic comet searches can be made using the standard techniques of wide angle and deep sky photography. Astrographs, Schmidt cameras, and even 35-mm cameras can be used. The key consideration in choosing a camera is focal length, balancing the need for a wide field against the need for sufficient image scale to recognize a comet. In addition, a 'fast' focal ratio (low f/number) helps in imaging a nebulous comet on film. Thus, Schmidt cameras are ideal for comet searching but fast astrographs and 35-mm or other format cameras with fast 135-mm to 500-mm lenses may be used. Electronic imaging, especially with CCDs (Section 3.7), has great potential when large format detectors are used. The major disadvantage of CCDs is the small area of sky covered when they are used with the focal lengths necessary to provide the image scale required to distinguish a comet.

Searches can be made anywhere in the sky, though photographers should consider the discussion in the visual procedures section (3.4.7) when designing a program. It is absolutely necessary to obtain two images of each field, separated by at least 45 minutes in time, to confirm not only that a suspicious object is real but also to determine its motion. Use a system such as that used by Eugene and Carolyn Shoemaker in their Palomar Asteroid and Comet Survey, described in Section 4.6.

As with other deep sky photography, high sensitivity and fine grain are desirable in the photographic emulsion. Black and white or color may be used to photograph to the sky limit (i.e., to the point that the background sky begins to fog the film). Color emulsions offer the chance of picking up a comet not only by its nebulous appearance but also because of its vivid blue coloration, even when in a field of galaxies (Mayer, private communication).

Once search photographs are obtained, the tedious process of examination begins. A magnifier, microscope, or slide projector may be used for careful, detailed examination of the images on the film. Use of a projection blink comparator (PROBLICOM) or stereo blink comparator may speed the search once reference slides centered on the same search fields are obtained (for details on these instruments see Section 4.6 and Mayer, 1977, 1978, and 1988, and Lazerson, 1984).

A photographic comet search has the potential of supplying important data immediately when a discovery is made. Competition from professional astronomers is formidable, adding to the challenge. Even if unsuccessful, the astrophotographer will build up a useful atlas of the sky.

3.4.11 Checking out suspects

The larger your telescope, the more nonstellar objects you will find and the more time you will need to spend checking out suspicious objects. Of the two ways to do this, the most common is simply to locate the object you see on some atlas, like Wil Tirion's *Sky Atlas 2000.0* (1981) or Tirion *et al.*'s *Uranometria 2000.0* (1987). Alternatively, if your telescope has an equatorial mount and a pair of setting circles, you can check the right ascension and declination of the object in which you are interested against a catalog of nonstellar objects, a good one being *NGC 2000.0* (Sinnott, 1988) from Sky Publishing Corporation.

Other sources to check are in preparation. Presently in production is a microfiche copy of the Palomar Observatory Sky Survey known as *Microsky*, being published by Deen Publications in Texas. The Space Telescope Science Institute in Baltimore, Maryland is preparing several versions of its guide star survey for distribution on CD-ROMs.

If your atlas or catalog shows no nonstellar object where there is clearly one in the field of your telescope, sketch its position and continue sweeping, going back in 15 minutes to check for motion.

3.4.12 Discovering a comet

If you hunt long enough, sooner or later you will find something that doesn't appear on an atlas. The first thing to remember is to stay calm, for the next minutes will be busy and important ones. Follow these steps:

(1) Does the object have a tail? Tapping the telescope lightly may help make a faint tail visible.

(2) Move the suspect around the field to check for the characteristic reflection that your object might be if you are near a bright star or a planet.

(3) If the object is faint, use high power to verify that the object is not a faint star or group of stars that might have appeared diffuse at low power.

(4) Sight along the tube so that you know the approximate direction in the sky. A Telrad Reflex Sight or Starbeam projection finder or similar 1x sighting device will help. Use an adjustable three prong star pointer (Canady, 1991) to find the area and relate it to your charts.

(5) Sight in the finder, and locate it more precisely on your charts.

(6) Check the position in a star atlas or catalog. David Levy finds that the Tirion *Sky Atlas 2000.0* shows over 70 percent of the suspects he finds in his 16-inch telescope. The *Uranometria 2000.0* atlas contains an even higher percentage of these suspects, but not all. But suppose the object is not shown, and its diffuse appearance still makes you suspicious.

(7) Draw a simple sketch, showing its position in relation to nearby field stars, including any useful alignments or mini-constellations in the field.

(8) When you next look at it minutes, or better, an hour later if possible, you should notice if it has moved. If there has been no motion by the time dawn or object-set has occurred, we advise strongly against sending a telegram. All comets do show motion eventually. You should wait to confirm motion, even if that process means you must wait another 24 hours.

(9) Where are the known comets in the sky? Check even the positions of comets that should be too faint for your telescope, although if you should be the first to notice an unexpected increase in a known comet's brightness, you definitely should report it. The positions of brighter comets are published in the major astronomy magazines, while IAU *Circulars*, the *Minor Planet Circulars*, and the International Comet Quarterly *Comet Handbook* are the good sources to check for known comets. If you subscribe to the computer service of the Central

Bureau for Astronomical Telegrams, you can ask their program to identify known comets or asteroids in the area around your suspect.

(10) If possible, have your sighting confirmed by an experienced and reputable comet observer. Tell that person the position, suspected nature, and direction of motion of the object, and ask him or her to confirm it.

(11) If you do detect motion and no previously known objects are reported to be in the area and your observation has been confirmed locally, it is now time to notify the Central Bureau for Astronomical Telegrams (CBAT) in Cambridge, Massachusetts. Its addresses are:

Telex: TWX 710–320–6842 ASTROGRAM CAM (Use this number 'for telegrams. If you address your wire to the Smithsonian Astrophysical Observatory, it could be delayed.)

E-mail: MARSDEN@CFA or GREEN@CFA (followed with one of these suffixes: .SPAN, .BITNET or .HARVARD.EDU) or EASYLINK 62794505.

They need the following information:

(A) Suspected nature of object.

(B) Right ascension and declination; although the CBAT uses equinox 2000 coordinates, the important thing is that you remember to state which equinox you are using.

(C) Direction and rate of motion. Preferably, this is supplied with a second position taken a half hour or more after the first. This motion is important, since it may be some time before a clear sky somewhere permits a central bureau observer to confirm your finding.

(D) The comet's magnitude.

(E) State whether the discovery was a visual or a photographic one; if you observed the object both ways, clearly distinguish how you made each aspect of your observation(s).

(F) Describe the object's appearance, including remarks about how condensed the object is, what its angular diameter is, and comment about the shape and length of the tail if you see one.

(G) The date and time of your observations, converted to Universal Time, or at least with time zone included.

(H) The instrument you used. Include the telescope's aperture, type, magnification, and $f/$ ratio for visual observations. If you are reporting a photographic or CCD find, add the film emulsion (or CCD type), the exposure time, the limiting stellar magnitude of the photo, and the size of the field in degrees or minutes of arc.

(I) Your name as discoverer, your address and telephone number.

Notes for new observers: Do not send a telegram unless you are certain that the object is a new comet. The philosophy of sending your announcement telegram, as some people have done, just to make sure you are the first, is hardly scientific and may embarrass you when you are informed that you have 'discovered' a well-known globular cluster. According to associate director Daniel W. E. Green of the CBAT, approximately 98 percent of comet discovery reports from unknown observers turn out to be spurious. Just because an object is fuzzy does not mean that it is a comet, and even if it is a comet it could be a known one. Galaxies, nebulae, and ghost images of bright stars are often reported as new comets. Fuzzy spots on single photographs should not be reported without confirmation.

If your discovery is confirmed by the IAU Central Bureau, then the comet will be named for you, unless others have also found it. Traditionally, comets are allowed a maximum of three names that represent the first three people to have seen and reported it. Therefore, you are in competition with other amateurs, professionals using photographic plates and CCD imaging systems, and even orbiting satellites.

3.5 Visual comet studies

Part of the fun of comet observing is in monitoring changes in individual comets, and in comparing the many comets visible each year with each other and those seen in previous years. There are a number of standard measurements that observers can make that permit easy comparison and can also contribute to cometary science.

Because of inherent variations in eyes, telescopes, and observing conditions, visual observations should be made with as many of those variables as possible minimized. Standardization in this manner will make data interpretation by yourself or others much easier. We recommend these methods. They are the best available based on current experience and analysis of comet observations.

3.5.1 Magnitude estimates

Visual photometry – brightness estimates – of a comet can have three targets: (1) the nuclear condensation, (2) the coma, and (3) the tail of the comet. It is very difficult to do accurate visual photometry on the tenuous, low contrast, filamentary tail of a comet. Photometry of the nuclear condensation is difficult because the true star-like nucleus is rarely seen, being invisible and/or almost always confused with a false 'photometric' nucleus or central condensation, and because it is seen against the bright background of the coma. Such photometry is also of questionable value, since it is strongly dependent upon aperture and magnification. However, a long, consistent series of magnitude estimates of the nuclear condensation made by the same observer with the same equipment (magnification, etc.) could potentially be quite useful. Observers should monitor the brightness of the central condensation and record the times of any major changes in brightness. These timings may prove very useful in the analysis of nucleus characteristics and provide data on its level of activity.

In the past, methods developed by N. Bobrovnikoff, J. B. Sidgwick, and M. Beyer have been used to estimate cometary brightness, specifically the coma's brightness. A study and comparison of these methods led C. S. Morris (1979, 1980) to suggest a new method. All these methods of visual photometry require the observer to memorize the image of the comet and then move the telescope to comparison stars that are probably some distance away.

With the Bobrovnikoff method, the observer selects several nearby comparison stars, some brighter and some fainter than the comet. Using a magnification of 1.5 to 2 power per centimeter of aperture (4 to 5 power per inch of aperture) to minimize the apparent size of the comet:

(1) The telescope is defocused until the comet and stars have a similar apparent size. The comet's apparent size will not change much when defocused but star images will.

(2) Go back and forth between a brighter and fainter pair of stars and interpolate the magnitude of the comet. (The interpolation method and example follow the description of the Morris method, below.)

(3) Repeat step 2 with several more star pairs.

(4) Take the average of the measurements in steps (2) and (3) and record it to the nearest 0.1 magnitude.

The Sidgwick or In–Out method is often used when a comet is too faint to withstand any defocusing:

(1) Memorize the 'average' brightness of the in-focus coma. This requires practice (unfortunately, this 'average' often varies among observers).

(2) Defocus a comparison star to the size of the in-focus coma.

(3) Compare the star's surface brightness with the memorized coma average brightness.

(4) Repeat steps (2) and (3) until a matching star is found or a reasonable interpolation can be made to the coma magnitude.

Although the Beyer method has fallen into disfavor because of the difficulty of using it and because of its sensitivity to background sky brightness, it still can be effective if used in a dark sky. The observer using this method defocuses the comet until it disappears against the sky and then searches for a star of known magnitude that disappears with the same amount of defocusing.

The Morris method matches the diameter of the moderately defocused comet with a defocused star. The procedure is:

(1) Defocus the comet's head to obtain an approximately uniform surface brightness across it.

(2) Memorize the image obtained in step (1).

(3) Match the comet image size with out-of-focus comparison stars. The stars will be more defocused than the comet.

(4) By comparing the surface brightness of the defocused stars and the memorized comet image, estimate the comet's magnitude.

(5) Repeat steps (1) through (4) until an accurate magnitude estimate to the nearest 0.1 magnitude is made.

When the comet's apparent magnitude is between that of two comparison stars, use the following standard interpolation method: Estimate the comet's difference from the brighter star in step sizes of tenths of the difference between the comparison stars. Then multiply the number of tenths by the magnitude difference of the stars and add this product to the magnitude of the brighter star. Rounded off to the nearest tenth, this is the comet's apparent magnitude.[2]

The AAVSO Variable Star Atlas and the separate individual variable star charts available from the American Association of Variable Star Observers are sources of comparison magnitudes. The atlas charts are useful only to about magnitude 8, and care must be exercised in using individual variable star charts due to their zero-point offsets from each other. For fainter magnitudes the North Polar Sequence should be used if no reliable comparison stars can be found near the comet [Fig. 3.10].

Magnitudes in *The AAVSO Variable Star Atlas* are given without a decimal point, and an underlined magnitude was determined photoelectrically. Thus, a star with 87 next to it is visual magnitude m_v 8.7, and a star with $\underline{33}$ next to it is photoelectric V magnitude 3.3. Magnitude estimates should be based on visual magnitudes, not V magnitudes, if possible, because a V magnitude is determined with an electronic photometer using a specified detector and glass V filter, which do not have the same response as the dark-adapted eye. At least three comparison stars should be used, but comparison stars that are obviously reddish should be avoided if they are distinguishable. Do not apply any corrections to the estimated magnitude. Record the value immediately after the estimate is made, along with the Universal Time to the nearest five minutes. Do not rely on your memory.

Always use the smallest possible instrument with the lowest magnification that lets you see the comet easily. Binoculars should always be used for comets brighter than magnitude 8 or 9, if possible. Using too large a telescope will

---

[2] Example: Suppose comparison stars *A* and *B* are magnitudes 7.5 and 8.2, respectively. Their magnitude difference is $8.2 - 7.5 = 0.7$. If the comet is 0.6 from *A* to *B*, then the estimated magnitude is $0.6 \times 0.7 + 7.5 = 0.42 + 7.5 = 7.92$ which rounds off to 7.9. A dark sky background will decrease the likelihood of underestimating the comet's brightness. Background sky brightness due to sources like city lights, natural airglow, twilight, the Milky Way, and the zodiacal light varies with position across the sky whether or not a dark observing site is used. Use extra care when making magnitude estimates when any of these background sources may affect your estimates by affecting the comet's contrast and apparent size against the sky.

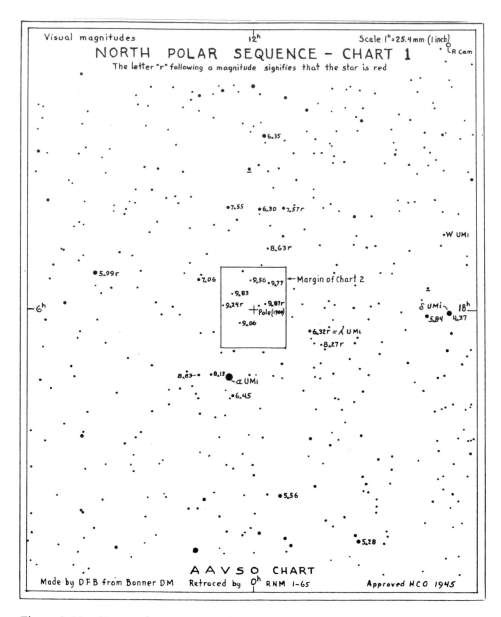

Figure 3.10   (See caption on page 52.)

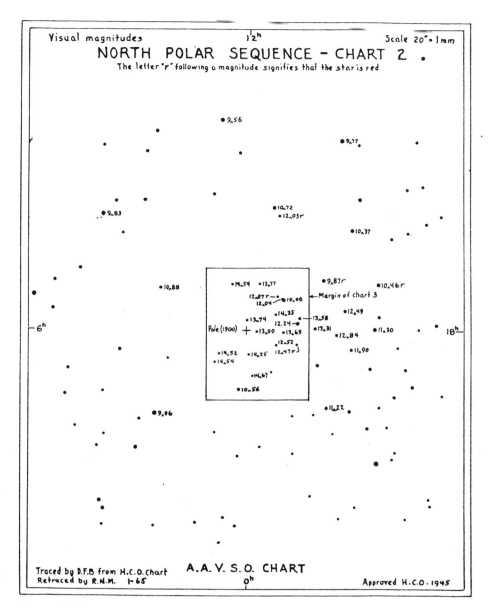

Figure 3.10 cont.

51

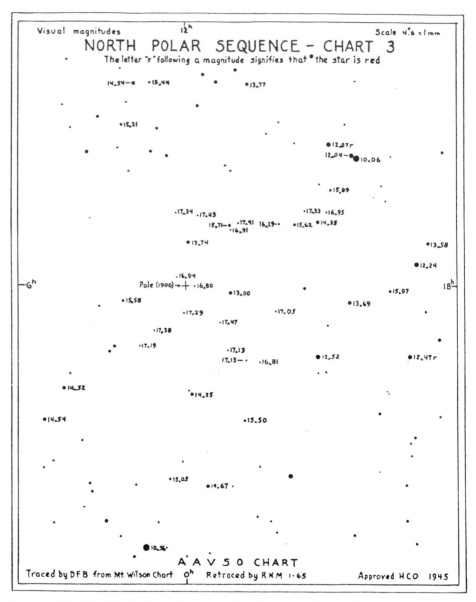

Figure 3.10   The North Polar Sequence of stellar magnitudes. These three charts show a wide range of magnitudes suitable for use in estimating the brightness of comets or asteroids, especially when no other suitable comparison sources are in the vicinity. Used with special permission from AAVSO.

result in errors to the estimates due to aperture effects that aren't fully known. The comet's distance also affects magnitude estimates. It is desirable to follow comets as long as possible, with the smallest instrument and lowest magnification possible. Thus, you may want to change telescopes during the comet's apparition. To maintain the continuity of your magnitude estimates and smooth the transition between instruments, you should make magnitude estimates using each of the instruments when a comet is within reach of them. A transition period should be used, even for changing only eyepieces.

Some observers find that using commercial 'comet' filters improves their observations of comets. Such filters may be useful in locating a comet (particularly those that aren't too dusty) and in seeing detail in the coma, but they should not be used for magnitude estimates.

3.5.2 Coma size measurement

Measuring the size of the coma and knowing the comet's distance (from an ephemeris) permits astronomers to know the true physical extent of a comet's head. Watching this development during a comet's apparition is interesting. In addition, the size of a comet's coma is useful in interpreting magnitude estimates. The same instrument should be used, and small apertures are again preferred.

One of four techniques should be used for coma diameter measurements. If the coma is elliptical, the length of both the long and short axes should be measured.

The simplest but least accurate method for measuring the diameter of the coma requires only an estimate of the coma size as a fraction of the separation of two stars. The angular separation S of the two stars is easily computed using their right ascensions (α_1 and α_2) and declinations (δ_1 and δ_2) in the formula

$$S = \cos^{-1} [\sin \delta_1 \, \sin \delta_2 + \cos \delta_1 \, \cos \delta_2 - \cos \, (\alpha_1 - \alpha_2)]. \tag{3.1}$$

The estimate should be made several times and the results averaged.

Another low-accuracy technique is to draw the coma on a detailed star atlas and measure its size using the scale of the atlas as a standard.

The coma diameter can be more accurately determined by using an illuminated cross-hair eyepiece (with a minimum of illumination) or an eyepiece having an occulting blade with the blade edge covering about half of the field of view.

First, orient one cross hair east–west so that a star drifts precisely along it when the telescope drive is off. Then simply time the interval necessary for the coma to pass by the north-south cross hair. The same technique works with an occulting-blade eyepiece once the blade edge is oriented north–south.

It is a good idea to begin with the coma out of the field of view when the clock drive is shut off. This will remove observational bias in the position of the coma's leading edge. The diameter, *d*, in minutes of arc can then be computed from the interval in seconds, *t*, and the comet's declination, *δ*, on the night of observation. The formula for this computation is

$$d = (1/4) \ t \cos \delta. \qquad (3.2)$$

We recommend that several interval measurements be made and the average value of these used.

   The third and most accurate method requires a reticle in the eyepiece or a filar micrometer. Once the scale is known, the angular diameter may be computed immediately from the measurement.[3] Eyepieces with built-in reticles are available from a number of manufacturers. Micrometer construction is discussed by Worley (1961) and by Polman (1977). Some are commercially available. Beish and Capen (1988) give detailed instructions in the use of a micrometer.

3.5.3 Degree of condensation

Degree of condensation (*DC*) provides a description of the coma's intensity profile (i.e., the change in brightness with distance along a diameter through

COMETARY BRIGHTNESS PROFILE

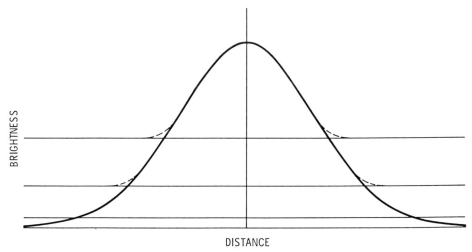

DISTANCE

Figure 3.11   A moderately condensed comet's intensity profile may look like this, with the contrast and apparent size decreasing if the background sky is bright.

---

[3] For the reticle, time the passage of the star along the length of the reticle, use Formula (3.2) to get the length in minutes of arc, and then divide by the number of reticle divisions to get minutes of arc per division.

the coma). It ranges from 0 (diffuse image, no condensation, flat, smooth profile) to 9 (star-like image with stellar intensity profile) [Figs 3.11, 3.12]. Occasionally, comets develop a coma with a sharp edge like a planetary disk. Experienced observers will generally rate this $DC = 9$ since the coma is not diffuse at all. It should also be noted that a condensed comet need not have a central condensation. (A central condensation is a distinct disk in the coma.) Morris (1981a) discusses DC in detail. Report DC as a whole number; if it really is between two numbers give the lower value followed by a slash (e.g., report a DC between 2 and 3 as 2/).

DEGREE OF CONDENSATION

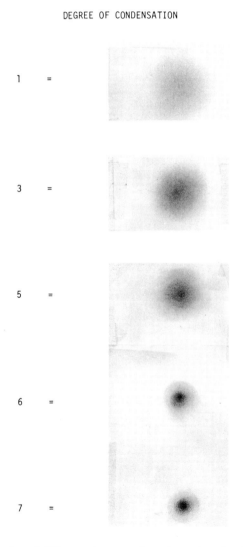

Figure 3.12   Examples of different degrees of condensation sketched by John Bortle. Courtesy of NASA/JPL/IHW.

Table 3.1. *Degree of Condensation*

DC	Description
0	Diffuse coma with uniform brightness, no condensation toward the center
3	Diffuse coma with brightness increasing gradually towards the center
6	Coma shows definite intensity peak at center
9	Coma appears stellar or sharp edged

3.5.4 Tail measurements

For tails less than 10 degrees in length, the tail's length should be estimated with respect to pairs of stars (as discussed for coma diameter observations). Do not use estimates made in terms of eyepiece or binocular fields-of-view. For longer tails, Equation (3.1) should be used with α_1, δ_1 referring to the comet's head and α_2, δ_2 referring to the end of the comet's tail. Background sky brightness can affect tail length estimates as well as coma magnitudes, as mentioned earlier. Take extra care when making these observations and make a comment in the notes section if there is a suspicion that the estimate was affected by background sky light [Fig. 3.13].

Position angle (*PA*) is best determined by accurately plotting the position of the head and tail on a detailed atlas and measuring the *PA* with a protractor. This method can be accurate to +/−5 degrees [Fig. 3.14].

PA can also be estimated by using the drift method to define east–west in an eyepiece and then estimating, to the nearest half hour, what time on an imaginary clock face the tail is pointing to. This method is only accurate to +/−15 degrees, which is half the angle between any two hour-marks on a clock. We do not recommend it for this reason.

A pointer attached to the outside of a cross-hair eyepiece can be used with a protractor or graduated piece of cardboard, wood, or metal fixed to the telescope and zeroed on north (with respect to the cross hairs and pointer). The cross-hair is rotated to the *PA* of the tail, and the value is read off the graduated circle. Due north is defined as 0 degrees *PA*, and *PA* increases through east, i.e., the trailing side of an object allowed to drift out of the field of view [Fig. 3.15].

When a star is seen through the tail, the *PA* can be calculated from the known positions of the star and the comet head. Use the formula

$$PA = \tan^{-1} \frac{\sin(\alpha_2 - \alpha_1)}{\tan\delta_2 \cos\delta_1 - \sin\delta_1 \cos(\alpha_2 - \alpha_1)} \quad (3.3)$$

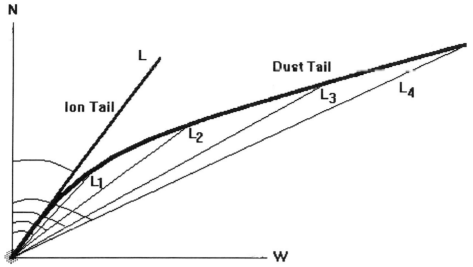

Figure 3.13 A comet's tails' lengths and position angles. Curved tails should be measured at several points. In this schematic drawing the ion tail has a length L and position angle different from the lengths and position angles measured for the curved dust tail. Note that the PAs for all the tails are between 270 and 360 degrees since the tails are in the northwest quadrant. If the angles indicated by the curved lines are measured, the corresponding position angles on the sky equal 360° minus the measured angle.

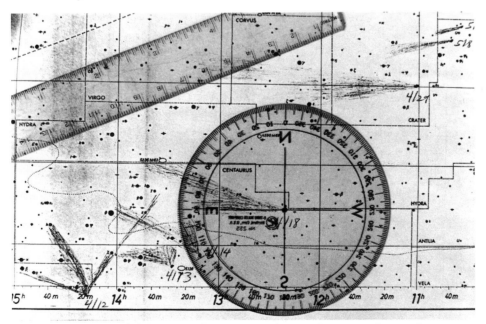

Figure 3.14 A careful drawing of a comet's tail(s) made on an atlas can be measured for position angle and length with a protractor and ruler. The ruler can also be used to extend a line along the tail's axis for measuring its position angle if the tail doesn't extend to the radius of the protractor. Note that the protractor is reversed because east and west on the sky are referred to geographic east and west. Drawings of Comet Halley by C. S. Morris.

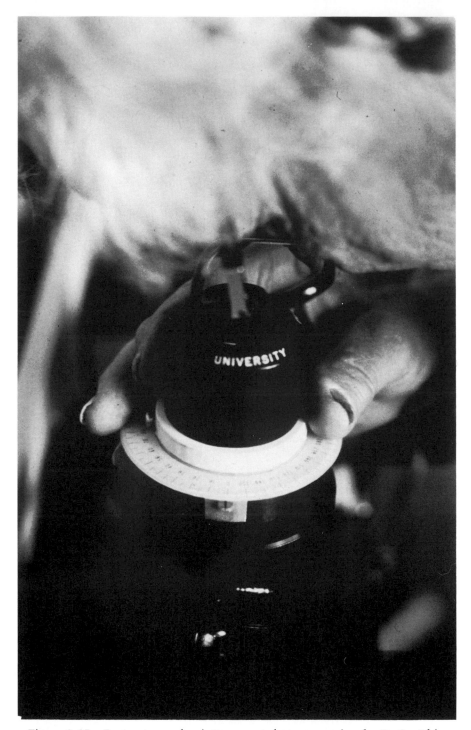

Figure 3.15   Protractor and pointer mounted on an eyepiece by R. Coutchie.

To determine the correct algebraic sign of the *PA*, determine the sign of [sin $(\alpha_1 - \alpha_2)$]. This will be the same as the sign of [sin *PA*]. That is,

$$\text{sign [sin } PA] = \text{sign [sin } (\alpha_1 - \alpha_2)]\qquad(3.4)$$

Short gas and dust tails are generally straight. The *PA* of long, curved dust tails should be measured at the tail root where it first leaves the coma and at various positions in the rest of the tail. A measurement of the distance from the nucleus for the *PA*-point measured should be included. This provides data on the curvature of the tail.

Notes on the presence or absence of structures in the dust tail are useful, as are reports on any changes seen and on the breadths of the gas and dust tail roots relative to the coma. A bright spine or dark 'shadow of the nucleus' (but not really a shadow) may be visible in the tail. Remember that tail observations are very sensitive to sky brightness and that moonlight or city lights can render such observations difficult. Twilight and the zodiacal light can also seriously affect tail observations.[4] Faint tails are sometimes more easily seen in large aperture instruments. A light tap on the telescope may improve a faint object's visibility.

3.5.5 Drawings

The principal data acquired with inner coma observations are best represented in a drawing. It is best to start by sketching the positions of field stars from a star atlas and then going to the telescope to fill in the cometary details. Observers can obtain valuable practice by drawing various nebular objects in the sky (see Eicher, 1983 and Romer, 1984).

Observations in twilight or moonlight can make fine detail in the coma visible and simply staring for a while at the comet allows more structure to be seen. Structures that may be visible include:

(1) Halos: circular, in whole or part, enhancements surrounding the central condensation.
(2) Fans: sectors of brighter material emanating from the central condensation.
(3) Rays or jets: radial features emanating from the central condensation.
(4) Envelopes: more than one level of brightness surrounding the central condensation, exhibiting a discontinuous change in stepping to the next level.
(5) Spines: a narrow, sharp, bright streak leading from the central condensation into the tail [Fig. 3.16].

---

[4] The times of the end of evening twilight and the beginning of morning twilight can be found in the *Astronomical Almanac*, RASC *Observer's Handbook*, *BAA Handbook*, and the *Sky-Gazer's Almanac* in the January *Sky and Telescope* for the current year.

Figure 3.16   A bright spine in the tail of Comet West on 27 March 1976.
Photograph by S. Edberg.

(6)   'Shadow' of the nucleus: not really a shadow, but a dark streak leading from the central condensation into the tail [Fig. 3.17].

(7)   Streamers: soft-edged, bright streaks seen in the coma and comet tails.

These should all be carefully drawn with their correct sizes, shapes, orientations, and positions with respect to the nuclear condensation. Don't rush. Use soft lead pencils or charcoal drawing supplies and paper stumps and erasers for smudging and erasing (available from stationery and art supply stores). Make 'negative' (dark comet on light sky) drawings of the comet.

Since high magnification is generally the best to use, long focal length refractors are superb for these observations. However, several magnifications or even several telescopes can be used for a very accurate representation of the coma and/or comet. The time of the drawing is an absolute necessity and the scale in minutes of arc/mm and the orientation with respect to north and east should be shown.

Measurements of the vertex distance and the semi latus rectum values or the pseudo latus rectum are very important to analysis of the nucleus. The vertex distance is the distance from the central brightness maximum to the vertex of the envelope as measured along the axis of the comet head. The

Figure 3.17   A 'shadow' of the nucleus in the tail of Comet Bennett. The comet was passing over a bright star when this picture was taken. Photograph by S. Edberg.

semi latus rectum values are both of the individual portions of the pseudo latus rectum. The pseudo latus rectum is a measure from the central brightness maximum to envelope edge perpendicular to the axis of the head, on both sides [Fig. 3.18].

The vertex distances and pseudo latus rectum or semi latus rectum values for internal envelopes should also be included. A written description of the features on the drawing is desirable.

High-quality drawings have value even with the advent of photography. The eye is very good at resolving fine detail during moments of good seeing and responds to an exceptionally wide range of intensity at one glance. True representation of the relative intensity and position of the nuclear condensation in the coma can yield valuable data on the state of excitation of the nucleus or of 'hot spots' on its surface. Determination of rotation and precession rates are possible using these data. Whipple (1981, in Marcus 1981, and private communication) urges amateurs to make drawings and measurements of haloes, envelopes, jets, and streamers observed in the coma. Good observations are extremely valuable for determination of various characteristics of the

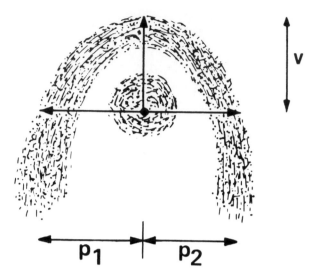

Figure 3.18 Depiction of a cometary coma with an envelope. V is the vertex distance. The two semi latus rectums, p_1 and p_2 may not be equal. In such a situation, they both should be identified and recorded separately, not totaled together. (Semi latus rectum is a mathematical term derived from a corruption of latus erectum, a Latin translation of the Greek orthia, for erect [B. Marsden, personal communication].) Courtesy of Dr. Fred Whipple.

nucleus. This is an area where real contributions to understanding comet nuclei are possible and where few observers are aiding the effort [Fig. 3.19].

3.5.6 Polarization studies

Visual observations of polarization in the coma may also be made. A polarizing filter attached to an eyepiece and then rotated may show the brightness of the coma to vary. A pointer and protractor (like those described earlier for measuring *PA*) can be used to determine the position angle of the filter which makes the coma brightest. Observers should be aware that some of the polarization they observe may be caused by the telescope optics. Look for polarization in a galaxy to see how much is generated by the telescope itself.

To find the correct position of the pointer follow this procedure [Fig. 3.20]:

(1) Lay down a piece of plain glass or plastic on the floor between yourself and a light source so that the light is reflected on the glass at a 45 degree angle.

(2) Looking at the light reflected in the glass, slowly rotate the filter. You will observe the reflection to have two minimums of intensity in one complete rotation.

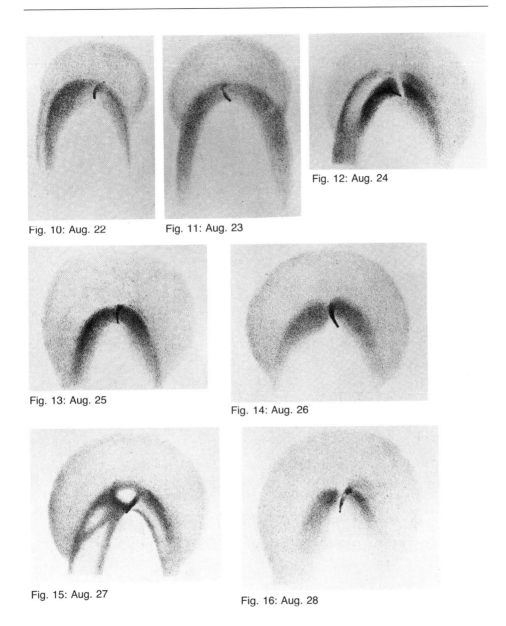

Fig. 10: Aug. 22   Fig. 11: Aug. 23

Fig. 12: Aug. 24

Fig. 13: Aug. 25

Fig. 14: Aug. 26

Fig. 15: Aug. 27

Fig. 16: Aug. 28

Figure 3.19   Sketches of P/Swift–Tuttle 1862 II made by J. F. J. Schmidt at the Athens Observatory.

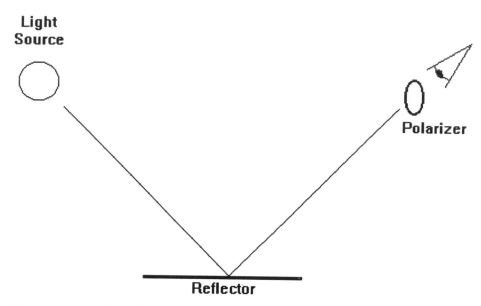

Figure 3.20   A piece of glass or plastic placed on the floor reflecting a light at about a 45 degree angle is suitable for determining the polarization direction of a polarizing filter as described in the text.

(3)   At either minimum, the pointer should be oriented perpendicular to the reflecting surface, i.e., it should point up (or down) if the glass is lying horizontally on the floor or on a table.

Visual comet observation is a good method for following the evolution of a comet, and comparing it to other comets. It yields valuable data and can be done by virtually any observer.

3.6 Photographic observations and electronic imaging

The increasing sophistication of amateur astronomers and their photographic equipment is an incentive to do cometary photography. Three areas of pursuit are available: general comet photography, astrometry, and spectroscopy can all be accomplished with instruments in amateur hands or obtainable by purchase or construction at home.

Electronic imaging with slow scan solid state detectors (CCDs, charge coupled devices) is a fast-growing field in amateur astronomy. A CCD is a highly efficient light-recording silicon chip whose light-sensitive pixels (short for picture elements) record the amount of light that strikes them. The picture, or image, that the chip gathers is then read by a computer. While these wondrous sensors do have some limitations, they are rapidly being overcome. Little has been published as yet by amateur astronomers using CCDs on comets but this sub-field is sure to blossom. Berry (1992 and 1991) is a good general source of information.

Before getting into the details of imaging comets certain details common to all these methods should be discussed.

3.6.1 Imaging technique

Ideally, a photographic emulsion should be very fine-grained to retain resolution to the seeing limit and very sensitive to decrease the data acquisition time. This combination has been the goal of film manufacturers for years. At this time, only the hypersensitizing of some emulsions allows adequate speed to be achieved. Otherwise, coarser-grained emulsions must be used for highest sensitivity, and fine-grained, less-sensitive emulsions are used in other situations.

Analogously, smaller pixels on CCDs permit higher resolution, and their sensitivity to light is legendary. A potential drawback to electronic sensors is their small active areas. Larger chips are becoming more affordable but care is still necessary in matching telescope focal length and detector for resolution and field of view.

Eastman Kodak's Technical Pan is the black and white (B/W) film of choice when hypersensitized because of its fine grain and high sensitivity. If hypersensitizing equipment is not available, high-speed (ISO 400 to 3200) moderately grainy films available from Kodak, Ilford, Agfa–Gaevert, or other firms should be used. In some situations nonhypered, moderate-speed films (ISO 100–125) with finer grain may prove useful including Technical Pan, Kodak Plus X Pan, Ilford FP4, and similar products.

Postexposure processing should be done as recommended by the film manufacturer. Low or normal contrast development should be used to bring out coma and tail details. The use of high-contrast developers like MWP-2, Kodak D-19, and similar types provide the necessary speed and contrast to show the overall extent of the comet.

Flat-fielding and dark current subtraction should be standard processing activities in any electronic imaging. Numerous image processing techniques can then be easily accomplished to get the most out of electronic images of comets.

3.6.2 Tracking a comet

A comet is a moving target and its tail structure is relatively faint. To record detail, long exposures will be necessary and correcting for the comet's apparent motion across the sky is absolutely vital. Even relatively short exposures will require accurate guiding on the comet since it is moving with respect to the stars.

There are several methods of correcting for a comet's motion across the sky, listed in order of decreasing accuracy [Fig. 3.21]:

(1) The most accurate method is to compute the comet's differential motions in right ascension and declination and then drive the telescope at those rates with respect to the clock drive. A microprocessor control makes this easy, but rate-calibrated drive correction motors on the telescope axes will also do the job. Arbour (1985) describes a stepper motor driven system mounted on a special plate at the tele-

CROSS-HAIRS

TAPERED POINTER

3. ON CONDENSATION

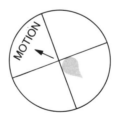

4. TANGENT CROSS-HAIRS

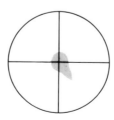

5. CROSS-HAIRS ON COMA

Figure 3.21 Methods (3), (4), and (5) for guiding a camera on a comet.

scope focus. Close to the horizon, differential refraction by the atmosphere can move the comet around the field of view in an irregular manner. Observers using this method should regularly check the tracking.

(2) Compute the angular motion and align a filar micrometer cross hair along the position angle of the motion. At predetermined intervals, offset the cross hair in the direction opposite the comet's motion and re-center the guide star. The correction rate depends on the angular rate of motion. This method is discussed in a pair of papers by Lines (1973 a,b). The first paper gives details on a micrometer microscope and the second one, and also Ridley (1984), give instruction on how to guide with one.

(3) Guide on any nucleus or distinct central condensation. Cross hairs or a tapered pointer are good for maintaining accurate centering.

(4) Compute the angular motion and align cross hairs tangent to the coma so that motion is directed along the diagonal of the opposite quadrant.

(5) The least accurate method is to center cross hairs on the coma and attempt to keep the cross hairs on the same point in the coma.

(6) Liller (1986) describes a method of tracking a comet by displacing the polar axis of the telescope a calculated amount. The offset depends on the observing site's latitude and the comet's distance from the meridian at the time of observation.

It should be obvious that off-axis guiding will not be effective since the photographic target – the comet itself – must be used to guide on in this case. Guiding methods (2), (3), (4), and (5) all require a co-aligned guide telescope.

3.6.3 General comet photography

Comet photography by amateurs can be classified in three categories: wide angle, moderate scale, and near-nucleus, reflecting the field of view (FOV) and/or target. The divisions separating these three categories are not sharp, but wide angle may be defined to mean the use of a FOV of tens of degrees and focal lengths of 100 mm or less with 35-mm film. By moderate scale a FOV of 5 to 10 degrees is implied; specifically Schmidt camera photography belongs in this category. Finally, near-nucleus photography implies focal lengths exceeding 1500 mm, with 2500 mm being an even better minimum useful value. Photographs taken at different times can be used for stereo imagery and, using photographic subtraction, studies of structural changes.

Such results are also possible with electronic images, with some of the processes taking considerably less time to accomplish. Bear in mind, though, that the smaller sizes of CCDs affect the choice of instrument focal length necessary to accomplish imaging at the scale desired.

The focal length used for a CCD (or even film of a different format) is directly proportional to the length of the diagonal across the sensor. Thus, a CCD with a diagonal length of 15 mm used with a telescope of focal length 1/3 of that used with 35-mm film will have the equivalent field of view since the film has a 44 mm diagonal length (44 divided by 15 is approximately 3).

3.6.3.1 Wide angle photography

This type of photography is especially suitable for bright comets with extensive tails. Remember to offset the aim of the camera so the head is near the end or corner of the frame so that more tail will be recorded. Doing this, though, means that the guide telescope is not aimed in exactly the same direction as the camera. For detailed studies of tail structure this is a disadvantage, so a view camera, or a system using some of the tilts and swings possible with a view camera, should be constructed to remove this disadvantage (Pansecchi, 1981).

Besides the beautiful and spectacular photographs possible, interesting results can be obtained with the use of color filters either during the observation or in the darkroom. The filters allow the photographic separation of the ion and dust tails if both are present. For tail photography, take a sequence including unfiltered, blue and orange images.

The ion tail can be isolated optically using an interference filter transmitting CO^+ wavelengths from 4100 Å to 4600 Å and excluding neutral molecular spectral lines and scattered solar continuum. The dust tail is separable using a filter transmitting a 'clean' continuum wavelength.

Suitable combinations of glass or gelatin filters, placed in front of the camera lens or immediately before the film plane, will allow tolerable isolation of the tail components. The blue filter's primary transmission band should be centered at 4400 Å with a width at the 50 percent transmission points of 900 Å and peak transmittance of at least 63 percent. The orange filter should have a sharp cut-on beginning transmission at 5400 Å and exceeding 90 percent transmission at wavelengths of 6500 Å and greater. Kodak gelatin filters 47A and 21, respectively, satisfy these specifications. The combination of 2B with 47A is even better in the blue. Glass filters matching these specifications are available from a variety of manufacturers – check with a camera store.

Blue images emphasize the ion tail of the comet while orange images emphasize the dust tail and its structure. A sequence (possibly a movie) made from these images through a comet's apparition may be very instructive. Several exposure sets each night will allow temporal changes to be detailed.

When the observing window for the comet is short (for example, when a comet rises shortly before twilight starts), it may not be possible to obtain two filter photographs of the tail. In such a situation, a color photograph can be used to record both tails simultaneously, with the advantage that observing conditions, guiding, and atmospheric refraction effects are identical. In the darkroom, photographic subtraction will allow isolation of the tails. Original color transparencies used for subtraction purposes should not be push processed because of possible color shifts during processing.

For the copying steps that are necessary in the method of photographic subtraction, the copies should have low contrast, even development, and equal image scales (avoid refocusing at different stages) for proper subtractions. Make an enlarged or contact B/W negative of the original color positive transparency, an enlarged or contact B/W negative of the transparency projected through a blue Wratten 47B filter, and an enlarged or contact B/W negative of the transparency through a red Wratten 25 filter. The two 'filtered' negatives should have comparable densities. From the filtered negatives, make B/W contact positives such that when a positive is placed emulsion-to-emulsion with its original negative, an even grey appearance over the comet and sky is produced (stars may not exactly cancel because they may be saturated).

To isolate the ion tail, make a print with the unfiltered negative and the 'red positive' facing each other emulsion-to-emulsion with stars aligned for overlap. To isolate the dust tail, make a print with the unfiltered negative and the 'blue positive' emulsion-to-emulsion [Fig. 3.22].

COMET KOHOUTEK 1973f

JAN. 11, 1974

| A. Blue ;-3800-4800Å | B. Red,-6000-6700Å | A minus B leaving CO⁺and CN emissions |

From color Ektachrome·EF   LPL

Figure 3.22   Photographic tail isolation of Comet Kohoutek. Photographs by S. Larson, Lunar and Planetary Laboratory.

3.6.3.2 Moderate scale photography

Much of what was said above regarding wide angle photography applies to moderate scale cometary photography. The observer can concentrate on the head and tail-root structure. For comets with long tails a complete image of the comet may have to be pieced together. There can then be problems with distortion at field edges making the fit of the montage parts difficult. Also, the images are not obtained concurrently. For the study of large faint comets or for comet searching this type of photography is unexcelled.

3.6.3.3 Near-nucleus photography

Narrow-angle coma photography can be accomplished with telescope apertures of 5 cm (2 inches) or more and focal lengths of 1500 mm or more (greater than 15 cm [6 inches] and 2500 mm are recommended). A series of bracketed exposures that vary by a constant factor (say, 2, so the exposure doubles from one to the next) will provide good results in spite of unknown amounts of atmospheric attenuation. A series will also better portray the full dynamic range of intensity from the inner coma to the outer coma as well as details such as jets, spiral structure, or small scale splitting of fragments from the nucleus [Fig. 3.23]. Photos obtained through polarizing filters (military surplus, not camera store types due to their inefficiency) at several known position angles may provide interesting data.

Figure 3.23   The asymmetric coma of Comet Bennett in 1970. Photograph by S. Edberg.

3.7 Conclusion

A variety of comet projects is available to the serious observer. They range from the simple to the complex and require skill levels from novice to expert. Astrophotographers looking for new challenges will find them in comets, which range from wandering deep sky objects to spectacular hairy stars visible in daylight.

4

Asteroids

4.1 Introduction

> The planet the little prince came from was scarcely any larger than a house!
> But that did not really surprise me much. I knew very well that in addition
> to the great planets – such as the Earth, Jupiter, Mars, Venus – to which we
> have given names, there are also hundreds of others, some of which are so
> small that one has a hard time seeing them through the telescope. . . .
> I have serious reason to believe that the planet from which the little prince
> came is the asteroid known as B-612.

In 1943 Antoine de Saint Exupéry published an enchanting tale called *The
Little Prince*. The story is about a prince who leaves his home on asteroid B-612
and visits Earth. Although his asteroid is small, it is rich with plant life, including runaway plants called baobabs which would take over the tiny planet,
splitting it into pieces if they were not removed every day.

On his way to Earth the prince visits asteroid 325, which is inhabited by a
king, and asteroid 328, on which lives a businessman. A lamplighter dominates
asteroid 329, lighting his lamp and then extinguishing it as the asteroid rotates
once each minute. Once the prince lands on planet Earth, he is astonished by
the huge planet's chorus of 111 kings, 900 000 businessmen, and 462 511
lamplighters. *The Little Prince* develops into a beautiful parable as prescient in
its human insights as it is forward-looking in its treatment of asteroids as
individual worlds. For when the book was published, asteroids were more seen
as inconvenient intruders on photographic plates than as objects of interest in
themselves.

That has all changed. Asteroids are now a wonderful field of research.
Among the unusual objects that have been discovered since 1990 are an asteroid locked near one of the Lagrangian points of Mars' orbit, asteroids which
grazed past us near the Moon's distance, an asteroid beyond Saturn's orbit,
and asteroids (or quiescent comet nuclei?) beyond Pluto's distance.

In a universe of large, mighty, and massive things, asteroids are more our
size. We can visualize an object in the sky the size of a state park, or a small
mountain; even though an asteroid may be 250 million miles away, it is in
the Earth's back yard and we can observe it with our own small telescopes
from our own back yards. These little rocks, once derided by astronomers as

tramps of our solar system, have inspired us in science and art. Generally referred to as asteroids, they are often called minor planets. They used to be called planetoids, a term more accurately descriptive than asteroids, but rarely used now.

Some science fiction writers and film makers see asteroid belts as terrible places where a spacecraft is certain to be destroyed. In reality this would rarely happen, as the several spacecraft that have safely made their way to the outer solar system can attest: there is a lot of empty space between the asteroids.

They also don't just exist in the large belt between Mars and Jupiter. In addition to that place, which we call the 'main belt', asteroids are found near the planet Jupiter and in orbits that intersect that of the Earth. Rather than being celestial tramps, these are a highly respectable group of objects.

4.2 Some historical notes

4.2.1 Origins

The field we now know as minor planets or asteroids (the name recommended by William Herschel) began not with observation but with a theory. It was a search for a new planet between Mars and Jupiter, begun not through a telescope but from the mind of Johann Daniel Titius of Wittenburg. Titius recognized the meticulous order represented by the planets at their increasing distances from the Sun, an order perfect except for an apparently missing planet between Mars and Jupiter. The numbers he came up with predicted the distances from the Sun of all the then-known planets so accurately that another astronomer, Johann Bode, embraced his theory enthusiastically around 1772. Bode's name, better known since he was Director of the Berlin Observatory and versed as a communicator, is now firmly attached to what is now known as the Titius–Bode Law. It mathematically defines the sizes of the orbits of the planets with a simple algorithm. To this day, and even with the breakdown of the 'law' for the outermost planets, science still does not understand why it works as well as it does.

The idea begins with 0 and then somewhat arbitrarily uses 3, doubling it over and over to get 6, 12, 24, 48, and 96. By adding 4 to each number, the resulting sequence becomes 4, 7, 10, 16, 28, 52, and 100. If one then divides each number by ten, one comes very close to having the correct distances, in astronomical units, of planets from the Sun. (An astronomical unit [AU] is simply the average distance of the Earth from the Sun.) In their average separations from the Sun, Mercury is 0.39 AU (close to 0.4), Venus 0.72 AU (0.7), Earth is defined as 1.0 AU, Mars is 1.52 AU (close to 1.6), Jupiter is accurate at 5.20 AU, and Saturn, at 9.54 AU, is close to 10.0. When Uranus was discovered a few years after the law was first publicized, its distance of 19.2 AU was so close to the value predicted by Bode's Law, 19.6, that the formula

73

seemed to be confirmed — with one 'sore-thumb' exception. There was no planet at 2.8 AU, and therein lies the law's historical importance: it inspired late eighteenth century astronomers to search for a planet at 2.8 AU from the Sun.

Actually, the numbers worked so well that the idea was touted as law, even though it was and is poorly understood. It remained only to find some dark deserted planet to fulfill the dream. In 1800 a group of astronomers calling themselves the 'Celestial Police' began a search for a faint new planet. Dividing the ecliptic into equal search regions for each member, the six observers scouted the sky for new worlds.

On January 1, 1801, Giuseppe Piazzi, a Sicilian monk searching independently of the Police, discovered an object whose orbit seemed to satisfy Bode's Law. However, this tiny planet, which needed a telescope to be observed, was far too small to be the one planet needed by the theory. The object was eventually called Ceres. A year later the Police scored their first victory when Heinrich Wilhelm Olbers, a prominent theoretician and asteroid and comet observer, discovered a second body that we now know as Pallas. Karl Harding, also a member, discovered Juno in 1804, and Olbers scored a second success with the discovery of Vesta in 1807. After several years with no further discoveries, the Celestial Police stopped their chase. There was a decades-long hiatus until 1845 when Karl Hencke found 5 Astraea.

The discovery of four of these tiny, faint bodies may have been perceived as a *reductio ad absurdum* disproval of Bode's Law. It was a real surprise. To preserve the law, some scientists speculated that there once may have been a large planet but that it had broken up. Most observers even then worried about that idea; a large planet would obviously have left many more substantial pieces. Maybe the pieces were there, waiting to be found.

At first, astronomers searched for asteroids visually. The procedure was simple, to compare what they saw through a telescope with what they could see on star charts. Chances are that a 'new star' would simply prove to be an error on the chart, a missing star, but when it actually moved from one night to another, the discoverer knew it must be an asteroid. Some 300 asteroids had been discovered in this tedious visual way before 1892, when the Heidelberg astronomer Max Wolf began using the techniques of photography to discover asteroids. (Wolf later became even better known for his recovery of Comet Halley in 1909.) The new procedure was a vast improvement. In taking photographs of a field of stars, the astronomer guided a telescope precisely to counteract the rate of Earth's rotation, so that any object that moves among the stars in the field of view would appear as a streak on the photographic plate. The thin streak would be but the beginning, however; afterwards a painful process of following and confirming the new object would ensue.

Once observers were picking up these minor planets on a regular basis, astronomers believed that they had the general orbits and nature of the aster-olds pretty well known. But this changed some fifty years ago when two important discoveries were made. In the first, Nicholas Dobrovnikoff found that asteroids have different colors – different spectral 'signatures'. This meant that not all asteroids are made alike.

Then in 1932, an asteroid was discovered with a most unusual orbit that took it closer to the Sun than Earth is. It was named Apollo after the god of the Sun. Five years later a group of German astronomers led by Karl Reinmuth were studying asteroids that they had photographed, and discovered an object moving very rapidly and only 400 000 miles from Earth. They called the object Hermes. Had this object, hardly larger than a hill, not been approaching Earth at the time of discovery, it would certainly have escaped notice, and Hermes has indeed been lost. Other examples of these Apollo asteroids, whose orbits cross that of the Earth, are Icarus, named by Walter Baade for the Greek hero who grazed the Sun, and Eros, the god of love. Yet another variety are the Amor objects, whose orbits cross that of Mars but not that of Earth, and the Atens, whose elliptical orbits are smaller than the Earth's but still cross it.

Those were the close asteroids. Early in this century, Max Wolf found the first of a group of asteroids so distant that they apparently 'shared' Jupiter's orbit, circling the Sun some 60 degrees ahead or behind the giant planet along the arc of Jupiter's orbit. Known as Trojan asteroids, they share the stable Lagrangian points 60 degrees ahead (L_4) or behind (L_5) Jupiter. In 1990 David Levy and Henry Holt discovered asteroid 5261 Eureka. It is the first *Martian* Trojan, in the L_5 region 60 degrees behind Mars in its orbit.

By March, 1971, asteroid studies had caught up with Antoine de Saint Exupéry's imagination. Interest in asteroids had become so widespread that a unique conference was held in Tucson. Organized by Tom Gehrels of the University of Arizona, the special thing about the meeting was that its only purpose was to present results about asteroids. Who but the imaginative storyteller Saint Exupéry could imagine that seventy different papers, each devoted to a slightly different area, could all be presented about asteroids!

With the dawn of respectability for asteroid science, research efforts multiplied. In 1977, Charles Kowal discovered Chiron, officially labelled as a minor planet (number 2060) but moving in a 51-year comet-like orbit entirely between the orbits of Jupiter and Saturn. In 1989 Karen Meech and Michael Belton detected a coma around Chiron. In 1992 a second Chiron-like object was found by the Spacewatch team on Kitt Peak, and independently by the team of Eugene and Carolyn Shoemaker and David Levy. The asteroid 1992 AD = 5145 Pholus follows a 93-year orbit inclined 25 degrees to the ecliptic and extending as close as 8.7 AU and as far as 32 AU from the Sun.

In 1990 the Galileo spacecraft made the first visit to an asteroid, Gaspra, showing it to have an irregular, cratered surface covered with a regolith of rock fragments, a violent past, and probable compositional variations across its surface [Fig. 4.1]. In December 1992 radar imagery of Earth-approacher

Figure 4.1   Minor Planet 951 Gaspra as imaged by the Galileo spacecraft in October 1990. Note the craters. Courtesy of NASA/JPL.

4179 Toutatis revealed an even more irregular shape and cratering [Fig. 4.2]. Galileo's successful 1993 flyby of Ida was closer than its flyby of Gaspra, and Ida is bigger. The images reveal a heavily cratered surface and strong relief [Fig. 4.3] and other measurements show compositional heterogeneity. A major surprise was the discovery of an apparent satellite 1.5 km (1 mile) across.

4.2.2 Discovery and naming of asteroids

By the end of 1991, more than five thousand asteroids had been discovered and numbered, with almost double that number discovered but waiting, with provisional designations, for enough observations to allow the calculation of good orbits.

How are asteroids named? In most cases asteroids are named by the discoverer, subject to ratification by Commission 20 of the International Astronomical Union. But the discoverer either has to observe his object sufficiently so that an orbit can be determined, or has to wait until someone else does that important work. Until then, the newly discovered object simply bears a provisional designation.

Giving asteroids provisional designations is not as simple a matter as naming their cousins, the comets. Each year is divided into 24 half-months, labelled

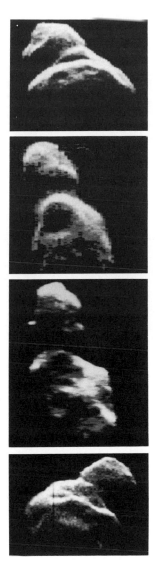

Figure 4.2   The contact binary asteroid 4179 Toutatis is composed of two objects, 4 and 2.5 km (2.5 and 1.6 miles) in average diameter. These images were made with the radar illumination from the top. Toutatis presents different faces during its rotation on the four days, 8, 9, 10, and 13 December 1992, these images were acquired. Note the irregular shape and rough surfaces, including craters. Courtesy of NASA/JPL.

Fig. 4.3   This picture of 243 Ida and its satellite was acquired by the Galileo spacecraft when it flew by the asteroid on 28 August 1993. The range of the spacecraft was 10870 km (6755 miles) and the resolution is about 100 meters (330 feet) per pixel. Note the many craters and irregular shape of the asteroid. Its 1.5 km (1 mile) companion is also irregularly shaped and is actually in the foreground in this view, closer to the spacecraft than Ida. Ida is an S class asteroid in the Koronis family. Courtesy NASA/JPL.

by the letters of the alphabet, excepting I (which could be mistaken for a roman numeral) and Z. During each half-month period (the 1st through 15th of each month, and the 16th to the end), a newly discovered asteroid is designated by the year of its discovery followed by two letters that indicate the half-month period and the number of discovery during that period. Once the designations in any half-month period have gone through the alphabet, the list repeats with a number; object 1990 UL3 signifies the fourth time through the alphabet during that half-month period.

Thus, 1990 MB is the second asteroid discovered during the twelfth two-week period (June) of 1990. 1990 is the year of discovery, M the half-month, and B designates the second discovery of that period. This procedure has changed just once, in 1925. Under the old rule, our asteroid would have been A990 MB, the A standing for asteroid.

Normally, one must observe a newly discovered asteroid for at least three and more often four oppositions (the time when an object is opposite the Sun in the sky), until its orbit is well understood. When the orbit is determined the asteroid is assigned a number, and the discoverer is then entitled to name it. An exception to this rule occurred when a major survey for asteroids was conducted in 1960 at Palomar Mountain in California and at Leiden Observatory in the Netherlands, and the newly found objects were given the designation P-L after their numbers.

While the discoverer has the right to name his or her asteroid, there are traditions and guidelines. The Trojans, positioned around the two Lagrangian points near Jupiter where the gravitational pulls of the Sun and Jupiter balance each other, are still named after Greek and Trojan war heroes. The Greeks are at L_4, preceding Jupiter, while the Trojans follow Jupiter at L_5. Each camp has a spy: 624 Hektor, a Trojan, orbits with the Greeks, and 617 Patroclus, a Greek, shares Trojan space.

If a discoverer has died, or has not shown any interest in naming an asteroid for at least ten years, others can propose a name which is then considered by the Director of the Minor Planet Center and Commission 20 of the International Astronomical Union. This committee may encourage name suggestions from people who had been associated either with the discoverer or with research on the asteroid.

Commission 20 strongly discourages names that are not in good taste, or names that are similar to those of other solar system objects. Normally two-word names or names over 16 characters long are out. Living political figures and controversial figures are usually not honored with asteroids, and a military victory less than a century old is considered inappropriate. Some names, controversial to some, may still be assigned if appropriate justification is presented: Asteroid 2807 is called Karlmarx and 1489 is Attila.

With 5000 minor planets numbered as of late 1991 (No. 5000 is called IAU for the International Astronomical Union), the named ones span a variety of interests. Asteroids have been named after people (2409 Chapman, 1877 Marsden, 4446 Carolyn, for planetary scientists Clark Chapman, Brian Marsden, and Carolyn Shoemaker). If a person's last name is already used, an asteroid might bear the combined forms of a person's first and last names (2068 Dangreen for Daniel Green of the Central Bureau for Astronomical Telegrams), or initials and claim to fame. Jay U. Gunter edited a long-lasting newsletter called *Tonight's Asteroids*; asteroid 2136 orbits the Sun as Jugta.

Although the commission now discourages this practice, there are a few asteroids named for pets. Asteroid 2309, Mr. Spock, was named for object 1971 QX1, discovered on August 16, 1971, by James Gibson at the Yale–Columbia Station, El Leoncito, Argentina. Touching and humorous, the official citation appeared in *Minor Planet Circular* 10 042 (1985 Sept 29):

> Named for the ginger short-haired tabby cat (1967–) who selected the discoverer and his soon-to-be-wife at a cat show in California and accompanied them to Connecticut, South Africa, and Argentina. At El Leoncito he provided

endless hours of amusement, brought home his trophies, dead or alive, and was a figure of interest to everyone who knew him. He was named after the character in the television program 'Star Trek' who was also imperturbable, logical, intelligent and had pointed ears.

One can name an asteroid after a tragic Shakespearean character, like 171 Ophelia, or a Greek goddess, like 55 Pandora. Some asteroids share names assigned to moons of planets (85 Io, 52 Europa), although this practice is no longer permitted, and obviously, two asteroids cannot share a name. Continents are noted in 1193 Africa, 916 America, 2404 Antarctica, 67 Asia; cities include 2118 Flagstaff and 2224 Tucson. Observatories are honored too, like 2322 Kitt Peak, 1951 Lick, and 2619 Skalnate Pleso. Concepts are noted in 679 Pax, and 2100 Ra-Shalom (named by Eleanor Helin to honor the Camp David peace accords; composers and writers are honored with 1815 Beethoven, 1814 Bach, 1818 Brahms, and 1034 Mozartia; 2106 Hugo, 3412 Kafka, 2625 Jack London, 2362 Mark Twain, 782 Montefiore, 2985 Shakespeare, and 2675 Tolkien. There is also 3131 Mason–Dixon, 2175 Andrea Doria, 1604 Tombaugh (the discoverer of Pluto) and 3310 Patsy (Clyde Tombaugh's wife and an accomplished artist). Teachers of astronomy are noted (3031 Houston, 3291 Dunlap, 3293 Rontaylor). The people who perished in the destruction of the space shuttle Challenger are remembered with 3350 Scobee, 3351 Smith, 3352 McAuliffe, 3353 Jarvis, 3354 McNair, 3355 Onizuka, and 3356 Resnik.

Directed by Brian G. Marsden at the Harvard–Smithsonian Center for Astrophysics, the Minor Planet Center is responsible for designating new finds of asteroids. To increase the efficiency of handling a large number of discoveries, the Center has introduced the pairing principle, which it recommends for observers reporting new asteroids.

A main advantage of this pairing principle is that it avoids the 'one night stands' where observers report a new asteroid observed on a single night and not observed again. Generally such observations are useless without sufficient follow-up work. Currently the Center will not accept or announce discoveries of any asteroid that has not been photographed, with accurate positions, on at least two nights.

Follow-up observations are vital if a newly found object is to have a well-understood orbit and not be lost. Thus, in the following moondark period (the nights between last and first quarter Moon when there are plenty of dark hours), observers are encouraged to submit at least two further nights of observations. The more observations, the better the object's orbit can be defined.

While this system has worked well, there have been problems regarding the announcement of rapidly moving near-Earth objects. If the rule is followed strictly, a single night observation of an asteroid moving quickly past the Earth and not followed up on a second night because of clouds, or the end of an observing run, could get lost as it zips by and fades away rapidly. Thus, an exception to this rule involves an asteroid like 1991 BA, which went very close to the Earth and was observed over a long arc of sky on a single night

The past few years have seen the recovery of some 'lost' asteroids. In 1991 878 Mildred, discovered by the famous Harvard astronomer Harlow Shapley and named for his daughter, was recovered after being lost for over 75 years. Although Shapley was no longer alive to enjoy it, his daughter Mildred was honored at a ceremony for the asteroid in Flagstaff. Now there is only one asteroid, 719 Albert, that has been numbered and which remains lost. Hermes, the Apollo object discovered in 1937, was not numbered and is now lost, its name not officially recognized.

4.3 The nature and origin of the asteroids

What are little planets made of? According to their inferred compositions, based on their spectra, we can divide them into at least two types. By far the most common is the 'C' type, resembling carbonaceous chondrite meteorites. These meteorites are made up of tiny millimeter-sized spheres of rock and carbonaceous material. A second type, 'S,' consists of about a sixth of the known objects, and these have silicate material, but also have a higher concentration of iron and nickel. A third type, 'M', appears to match meteorites made almost entirely of nickel and iron, without any silicates. A subgroup of 'M' consists of metals apparently embedded in silicates.

We have two basic concepts concerning the present structure of the asteroids. One proposes that they are chunks of material that have been formed into random shapes over billions of years as they collided with their cohorts. Gaspra (Fig. 4.1) is an example of this concept.

In a second model, they are battered heaps of rubble, held together by their own gravity but so weak that they assume shapes dictated by the laws of fluid mechanics. It is possible that asteroids have shapes similar to those of fluid bodies in equilibrium.

This second idea is not new. We know that large planets behave as fluid bodies since the pressures throughout their interiors, caused by gravitational compression, cause their material to flow like a viscous fluid. To a large extent, the shape of such fluid bodies depends only on their rotation rate and density. A body that does not rotate at all would be a sphere. Bodies like Jupiter and Saturn rotate fast enough so that they are flattened at the poles and bulging at the equator. We call this shape a 'Maclaurin spheroid'. On the other hand, bodies that rotate much faster than that assume strange spheroidal shapes with three axes of different lengths – we refer to these as 'Jacobi ellipsoids'.

Even though the asteroids are so much smaller, and obviously not fluid, some of them also may have shapes like fluid bodies.

But how can an asteroid behave like a fluid body? Imagine breaking up some hard, well-packed sand repeatedly until it flows so freely that it almost

81

assumes the properties of a fluid. The environment of the asteroid belt is somewhat similar. With thousands of asteroids in the main belt between Mars and Jupiter, a large number of collisions, spread out over the life of the solar system, are inevitable. Considering the general trend and direction of their motion, asteroids typically smash into each other at a speed of about five kilometers per second.

At that rate, small asteroids would disintegrate, leaving craters in the large targets they hit; the Galileo spacecraft showed asteroid 951 Gaspra as a body covered with craters of different sizes, the results of a long history of encounters with smaller objects. But if the smaller asteroid approaches the size of the larger one, the size of the crater becomes very large. For example, Voyager 2 photographed a huge crater (subsequently named Herschel), covering a large area of Saturn's moon Mimas. If the two colliding asteroids are almost the same size, one or both might simply be shattered into many fragments. The punch line: if the shattered asteroid is a small one, the pieces, or ejecta, become tiny asteroids orbiting the Sun in their own orbits.

Inside the main belt there are clusters – called families – of asteroids that are believed to be the remains of large asteroids that were destroyed long ago. However, it is not asteroids themselves that are close together in space. Rather, some of their orbital elements – their semi-major axes, their eccentricities, and their inclinations – define them as being part of the same original body. Named after their lowest-numbered members, three such families are known as Eos, Koronis, and Themis. Altogether there are more than 100 such families, each believed to be the result of a collision in primordial times.

If a large asteroid breaks up after a collision, then the fragments may not have enough energy to escape each other and instead they simply collapse back into a single body. In such a fashion an initially coherent body can be converted into what we call a 'gravitationally bound rubble pile'. If collisions happen again and again over billions of years, the body would repeatedly break up and reassemble, assuming, like the ball of sand on a beach, a shape like that of a fluid body.

How were the asteroids formed? Were they really the remains of a broken-up planet or were they material that might have formed a planet and did not, as was commonly thought a generation ago? According to current thinking, when the primordial solar nebula cooled, material of different composition condensed at varying distances from the center. The regular chondrites condensed at the inner edge of the asteroid belt, and the carbonaceous material consolidated at cooler temperatures further out in the belt. Much further out, the ices condensed to form independent comets or they congealed into the huge planets.

There are other places besides the main belt where asteroids can remain throughout the life of the solar system. Asteroids are found in the 'Trojan' zones near Jupiter and Mars and could be stable in those zones of other planets, an area of ongoing discussion and observation. In the spaces between other major planets, small bodies are unlikely to be gravitationally perturbed out of their orbits. Perhaps Chiron and Pholus are eccentric members of another belt. Even the space between Mercury and the Sun may host an asteroid belt.

Why didn't the asteroids in the main belt accrete to form some small, Mars-sized planet? As Shakespeare did not quite write, 'The fault, dear Brutus, is not in themselves, but in Jupiter, that they are underlings.' That monstrous planet's gravity may have caused perturbations or, by some other mechanism, kept the planet-to-be from forming. Instead, the orbits of the small bodies were changed, so that instead of gradually coalescing into a large body, they collided into each other, breaking themselves apart.

Some scientists do not believe that Jupiter's gravity was totally responsible for the disruption. Instead, they have suggested that the orbit of another proto-planet, perhaps the size of Mars or Earth, was perturbed by Jupiter into the asteroid region. It quickly disrupted the small protobodies and then, they argue, it was thrown out of the solar system entirely by a second close encounter with Jupiter. In its wake, the protoplanet left a group of thousands of tiny objects colliding into each other.

4.3.1 The distribution of asteroids in space

Imagine a strange football field laid out in space with the Sun at one goalpost and the orbit of Jupiter at the other. It is divided into zones and regions whose borders crisscross each other.

Across this field, zones appear as parallel 'yard lines' that mark areas where an object would orbit the Sun in a commensurate period relative to that of Jupiter. These lines are called 'zones of commensurability'. Groups of asteroids in one of these zones can orbit the Sun in a precise fraction of the amount of time it takes Jupiter to complete one of its orbits. For example, the Jupiter Trojans are in a 1:1 zone of commensurability; they orbit in the same time it takes Jupiter to orbit.

Other zones of commensurability house groups of safely orbiting asteroids, like the 3:2 and the 4:3 zones.

But to make the game more interesting, some zones of commensurability are illegal. Any asteroid that somehow falls into one of these zones will have an orbit that is unstable. Examples are the 4:1 commensurability, where an object would orbit precisely four times faster than Jupiter does, as well as 3:1, 5:2, and 2:1, and others.

But this unusual game has other prohibited regions. These are called secular resonances, each given the Greek letter ν followed by a number. An asteroid is located in a secular resonance if the rate of precession of its orbit's longitude of nodes, or longitude of perihelion, is a small rational fraction of that of one of the major planets, notably Jupiter. Precession refers to a slow angular change in the orientation of the orbit. It is referred to the longitude of the nodes, which is the angle between the position of the vernal equinox and the position of the ascending node, or the longitude of perihelion, which is the angle between the position of the vernal equinox and the position of the body's perihelion.

There are no winners or losers; the only object for the players, the asteroids, is to stay on the field by avoiding either the zones of unstable commensurability or the secular resonances. Should an asteroid's orbit, slowly changing due to the influences of all the planets, cause it to stray into either of these types of forbidden areas, its orbit would be rapidly perturbed. It could even be expelled from the game, perhaps by being injected into a new orbit that would take it close to Jupiter. When that happens, the asteroid could be thrown out from the solar system (and the game) altogether. But the other possibility is that the new orbit will fling the asteroid into the inner solar system. If that happens, sooner or later the asteroid would collide with one of the planets. Alternatively, a near-encounter with an inner planet might return the asteroid to the neighborhood of Jupiter. Then the prodigal asteroid could be ejected from the solar system.

Between the zones of commensurability and the secular resonances lie areas of stability in which asteroids orbit the Sun. By far the majority of known asteroids exist in the main belt, a large 'safe area' which stretches between 2.0 and 3.7 astronomical units from the Sun. (The Earth orbits at one astronomical unit from the Sun, Jupiter at 5.203 AU. Technically, the main belt is bounded on the sunward side at 2.0 AU by secular resonances $\nu16$ and $\nu6$, as well as the 4:1 commensurability.) The 'illegal' and nearly vacant 3:1, 5:2, and 2:1 commensurabilities, which are called Kirkwood gaps after their discoverer, are divisions in the main belt.

Now we can look at the playing field to visualize where the various groups of asteroids lie. For example, the Hungarias (a group of asteroids named after the first one discovered, 434 Hungaria) have orbits whose mean semi-major axes lie at about 1.9 AU. The Phocaeas, a group named for 25 Phocaea, are on the other side of the $\nu16$ secular resonance, ranging from about 2.1 to 2.4 AU. The Phocaea region is bounded by the $\nu16$, $\nu5$, and $\nu6$ secular resonances and by the 3:1 commensurablity ratio. Both the Hungarias and the Phocaeas have orbits that are inclined as high as 25 degrees from the plane of the ecliptic. Thus, our unusual football field is not flat but three-dimensional. The asteroids in the small region of stability that lies beyond the 2:1 resonance are referred to as Cybeles.

The Hildas, Thules, and Trojans play the game by special rules. The Hilda orbits go through oscillations, although their mean semi-major axes are precisely on the resonances 3:2 for Hildas, 4:3 for the Thules, and 1:1 for the huge class of Trojans.

4.3.2 Near-earth asteroids

When Apollo was discovered in 1932 and Hermes in 1937, it was thought that these asteroids were rare oddities. But in recent years the search for small objects that approach the Earth has become an important aspect of asteroid studies. Throughout much of the second half of the 1980s, a number of programs in the United States and Japan have discovered a large group of Apollo asteroids whose orbits cross that of the Earth, Amor asteroids whose orbits cross that of Mars, and Atens, whose orbits are *within* that of the Earth.

Besides the scientific interest, there is the potential for practical benefit from these searches. One benefit is for the robotic and/or manned exploration and perhaps utilization of these objects. The other is prophetic: if one of these asteroids hit the Earth (and virtually all the Apollos will some day) the result would be catastrophic. With the increasing interest in the theory that asteroids or comets might have caused the extinctions of whole families of life in the past, some groups have been organized to search for Earth-approaching asteroids.

According to Eugene Shoemaker, an Earth-crossing asteroid is an asteroid whose orbit will intersect the Earth's orbit as a result of long-range gravitational perturbations. He estimates the population of asteroids so defined as about 2000, with a diameter of 1 km or greater. At present, about a tenth of this number have been discovered.

To learn about the population of asteroids that might strike the Earth some day would seem to be a vital scientific project, even though funding levels for the three professional search programs currently running in the United States are very low. Eleanor Helin conducts one such program. Her 'Planet-Crossing Asteroid Survey' uses the 18-inch (45-cm) Schmidt camera at Palomar Mountain. With a small group of assistants, in six nights of observing each month the team photographs and scans a substantial portion of the sky. Depending on the weather, during a good observing run they might take three hundred films or more, with each film repeated after a short period. The film pairs are scanned with an instrument called a stereomicroscope, through which any moving objects appear three-dimensionally above or below the level of the background stars. Depending on which order the films are placed on the machine, objects moving to the west, or retrograde, can be made to appear higher than background level, while objects moving to the east, or prograde, appear lower.

In the summer of 1989, David Levy joined the second search team, that of Carolyn and Eugene Shoemaker and Henry Holt. They also observe at Palomar Mountain Observatory's 18-inch Schmidt in southern California, where for a period of seven nights each month they take as many as 60 films per night. During a typical year the Shoemakers conduct the observing for eight months while Holt handles the remaining observing runs.

Searching for asteroids and comets the Shoemaker way is strenuous business. On a typical night, Eugene Shoemaker and Levy might be sharing the observing while Carolyn Shoemaker scans the previous night's films. With the telescope horizontal, Levy will present a film holder with loaded film to Shoemaker, who installs it into its place at the telescope's focal point. Shoemaker adjusts the focusing wheel, removes the film holder cover and closes the telescope's sliding doors. After a five-second countdown, Shoemaker opens the shutter and begins the exposure. Eight minutes later the exposure is over, the film is removed from the telescope, and the procedure is repeated. A set of films usually covers four, sometimes five, fields of about 8 degrees in diameter; after the four fields are photographed, they are photographed again. Ideally, 45 minutes of time separate exposures of the same field.

Carolyn Shoemaker's major task is examining the films for asteroids, comets, and variable stars. Having a single member of the team do all the scanning is a procedure that has proven effective in many other programs, notably the 14-year planet hunting effort at Lowell Observatory, during which several observers took photographic plates but only one, Clyde Tombaugh, scanned them all. Although Tombaugh was not specially searching for asteroids he found more than 700 new ones.

Spacewatch is the name of a third team based at the University of Arizona in Tucson. Using a 90-cm (36-inch) reflector and CCD at Steward Observatory on Kitt Peak, the *Spacewatch* team of Tom Gehrels, Jim Scotti, and Robert Jedicke try to discover asteroids that are approaching Earth. With the slogan 'Find them before they find us', the *Spacewatch* team has been exploring (they call it 'scanning') strips of sky since 1983. The ability of the system to examine areas of sky down to fainter than 20th magnitude, much fainter than photographic surveys, has allowed it to find some very interesting asteroids. For example, 1991 BA is an object some ten meters across that whisked by the Earth at less than half the distance from the Earth to the Moon. Because of the small area scanned by the *Spacewatch* telescope, a reasonable projection suggests that as many as 50 of these tiny asteroids pass between Earth and Moon each day.

When the telescope is in the scanning mode, the drive is usually off, letting the Earth's rotation do the work of moving the sky past the telescope's detector. Each scan is repeated, so that there is about a half hour difference in time between the images, allowing a new asteroid to move noticeably between one scan and the other.

4.3.3 Asteroids and comets

Comets and asteroids may have more in common than simply being members of our solar system. When a comet coming in from a great distance is gravitationally perturbed by a planet like Jupiter, its orbit may gradually change so that it travels around the Sun in less than 200 years. After thousands of these orbits, much of a comet's ices have sublimated away, leaving just the rocky core. Alternatively, the rocky crust may be so thick that the ices inside can no longer sublimate. Asteroids 2201 Oljato and 3200 Phaethon may be examples of defunct comets. Oljato may be surrounded by some sort of debris, and Phaethon, discovered in 1983 by IRAS, the Infrared Astronomical Satellite, appears to be the parent of the Geminid meteor shower and shows no cometary gas emissions to the limit of telescopic detectability.

Just how closely a comet can resemble an asteroid was made clear to David Levy after the discovery of an object in November 1989 by the team of Carolyn and Eugene Shoemaker and Levy. Designated as asteroid 1990 UL3, this object turned out be in an elongated orbit taking it out to the distance of Jupiter, an orbit more typical of a comet than of an asteroid. Could a telescope and CCD detector far more powerful than the 18-inch Schmidt camera and its films (that were used in the discovery) reveal that the asteroid is really a comet?

In December 1990 Steve Larson and Levy, as part of a long-running program to take detailed images of as many comets as possible, decided to take a close look at 1990 UL3 to see if it showed any signs of cometary activity. Our procedure was to take a number of five-minute CCD exposures with the 152-cm (61-inch) Mount Bigelow telescope near Tucson. Then we would electronically add them together, an approach that is really the electronic way to make composite prints in the darkroom using several negatives. The result decreases the significance of the noise inherent in the imaging method. Each image would be flat-fielded, which removes the effects of any unevenness in the CCD chip. Also, by displaying all five images one after another, the asteroid would give its presence away by appearing in a slightly different position in each frame as it moved slowly. Then we would study the asteroid carefully to see if it had any trace of fuzziness around it.

We didn't have to do all that. As soon as we looked at the first flat-fielded image, one of the faintest 'stars' showed an even fainter tail stretching for about 30 seconds of arc, or 60 000 km in space. After several more confirming images over two nights, asteroid 1990 UL3 was officially announced by the Central Bureau for Astronomical Telegrams as Periodic Comet Shoemaker–Levy 2 after the original discoverers. The faintness of the tail relative to the head indicated that the comet may possibly have used up most of its volatile material, and that eventually it might cease to show any tail or coma at all, to then appear as an asteroid.

More recently, minor planet 1979 VA, after receiving its permanent number of 4015, had its orbit extrapolated back by Ted Bowell. He found 1949 images of it on Palomar Sky Survey plates that showed a tail. 4015 has now been identified as Comet P/Wilson–Harrington, 1949 III, which has perhaps given up its last gasps as a comet.

Another recent discovery is that of a tail, to go with the coma discovered a few years ago, associated with 2060 Chiron. While Chiron is showing all the classic cometary phenomena a few years before perihelion and outside Saturn's orbital distance, 5145 Pholus is near perihelion inside Saturn's orbital distance and is not showing any cometary activity.

4.3.4 The future

We have only the barest glimpse of what the Earth's immediate neighborhood looks like. In a sense the Earth is travelling down a highway with a very poor windshield. With increasing interest in the possibility that Earth might some-day be clobbered by a passing asteroid, as this book goes to press the United States Congress is considering a proposal to set up a series of CCD-equipped wide-field telescopes so that a catalog of all asteroids of 1 km diameter or larger would be 90 percent complete within the next 25 years. Such a program would increase vastly the discovery rates of all asteroids, and would give us a more comfortable view of the highway around us. Ideally, we would then know about an asteroid about to hit us, allowing for time to develop a means to deflect it from our path.

4.4 Observing asteroids

4.4.1 Starting your observing program

Following an asteroid as it moves among the stars as the nights go by is a pleasant experience that grows on you. David Levy started out by viewing only the brightest asteroids during the 1960s with a 20-cm (8-inch) reflector while Stephen Edberg used a 25-cm (10-inch) reflector. A telescope or a good pair of binoculars, a set of predicted positions, a dose of enthusiasm, and a lot of patience is all you need to begin a successful career of observing asteroids.

4.4.2 A life-list

The idea of a life-list began with Roger Tory Peterson's *A Field Guide to the Birds*.[1] It is a record designed to encourage amateur bird-watchers to search for different species of birds, and it should work just as well with asteroids. The object is to record the date on which you first see an asteroid. In Table 4 we have included the first 142 asteroids, as well as some others that can reach 12th magnitude or brighter at their periods of opposition to the Sun. Since the magnitudes are 'B' (for blue) filter magnitudes the observed brightnesses could be about a magnitude brighter when they are seen visually. But, more important, these magnitudes are the optimum brightnesses of the asteroids at

---

[1] Peterson, R. T., *A Field Guide to the Birds*, Cambridge, Houghton Mifflin, 1934, 1939, 1947.

their very best, when their elliptical orbits bring them closest to the Earth and in opposition to the Sun. So, in most circumstances, they are bound to be fainter than listed here.

Table 4.1. *Life-list of asteroids*

Asteroid	$M_{(B)}$	Date	Comments
1 Ceres	7.2		
2 Pallas	6.8		
3 Juno	7.8		
4 Vesta	6.2		
5 Astraea	10.0		
6 Hebe	8.1		
7 Iris	7.5		
8 Flora	8.6		
9 Metis	9.2		
10 Hygiea	9.7		
11 Parthenope	9.8		
12 Victoria	9.2		
13 Egeria	10.0		
14 Irene	9.5		
15 Eunomia	8.3		
16 Psyche	9.8		
17 Thetis	10.8		
18 Melpomene	8.2		
19 Fortuna	9.8		
20 Massalia	9.2		
21 Lutetia	10.1		
22 Kalliope	10.6		
23 Thalia	9.4		
24 Themis	11.2		
25 Phocaea	9.8		
26 Proserpina	11.4		
27 Euterpe	9.4		
28 Bellona	10.7		
29 Amphitrite	9.6		
30 Urania	10.3		
31 Euphrosyne	10.4		
32 Pomona	11.1		
33 Polyhymnia	10.6		
34 Circe	12.0		

Table 4.1 *(cont.)*

Asteroid	$M_{(B)}$	Date	Comments
35 Leukothea	12.1		
36 Atalante	10.6		
37 Fides	10.2		
38 Leda	11.9		
39 Laetitia	10.2		
40 Harmonia	10.1		
41 Daphne	9.7		
42 Isis	9.6		
43 Ariadne	9.9		
44 Nysa	9.6		
45 Eugenia	11.0		
46 Hestia	11.1		
47 Aglaja	11.7		
48 Doris	11.6		
49 Pales	11.2		
50 Virginia	11.2		
51 Nemausa	10.5		
52 Europa	10.7		
53 Kalypso	11.5		
54 Alexandra	10.6		
55 Pandora	11.0		
56 Melete	10.7		
57 Mnemosyne	11.6		
58 Concordia	12.8		
59 Elpis	11.3		
60 Echo	11.0		
61 Danaë	11.6		
62 Eratio	12.5		
63 Ausonia	10.4		
64 Angelina	11.1		
65 Cybele	11.7		
66 Maja	12.4		
67 Asia	10.7		
68 Leto	10.2		
69 Hesperia	10.7		
70 Panopaea	10.9		
71 Niobe	10.6		
72 Feronia	11.4		
73 Klytia	12.5		
74 Galatea	11.5		

Table 4.1 *(cont.)*

Asteroid	$M_{(B)}$	Date	Comments
75 Eurydike	10.9		
76 Freia	12.4		
77 Frigga	11.9		
78 Diana	10.9		
79 Eurynome	10.4		
80 Sappho	10.0		
81 Terpsichore	11.7		
82 Alkmene	11.1		
83 Beatrix	11.7		
84 Klio	11.1		
85 Io	10.5		
86 Semele	12.3		
87 Sylvia	12.0		
88 Thisbe	10.4		
89 Julia	9.5		
90 Antiope	12.3		
91 Aegina	12.2		
92 Undina	11.2		
93 Minerva	11.0		
94 Aurora	12.1		
95 Arethusa	12.0		
96 Aegle	11.7		
97 Klotho	9.8		
98 Ianthe	11.9		
99 Dike	12.3		
100 Hecate	11.8		
101 Helena	11.3		
102 Miriam	11.6		
103 Hera	11.5		
104 Klimene	12.5		
105 Artemis	10.6		
106 Dione	11.6		
107 Camilla	12.0		
109 Felicitas	11.0		
110 Lydia	11.6		
111 Ate	11.4		
112 Iphigenia	12.6		
113 Amalthea	11.6		
114 Kassandra	11.7		
115 Thyra	9.7		

Table 4.1 *(cont.)*

Asteroid	$M_{(B)}$	Date	Comments
116 Sirona	11.4		
117 Lomia	12.6		
118 Peitho	11.4		
119 Althaea	12.1		
120 Lachesis	12.3		
121 Hermione	12.1		
122 Gerda	12.8		
123 Brunhild	12.5		
124 Alkeste	11.9		
125 Liberatrix	12.6		
126 Velleda	12.3		
127 Johanna	12.5		
128 Nemesis	11.0		
129 Antigone	10.3		
130 Elektra	10.7		
131 Vala	13.6		
132 Aethra	10.1		
133 Cyrene	12.5		
134 Sophrosyne	11.8		
135 Hertha	10.3		
136 Austria	12.5		
137 Meliboea	11.6		
138 Tolosa	11.4		
139 Juewa	11.1		
140 Siwa	11.3		
141 Lumen	11.3		
142 Polana	12.9		

At this point we no longer include each numbered asteroid, only a selection.

216 Kleopatra	12.1		
233 Asterope	11.8		
234 Barbara	11.0		
236 Honoria	11.6		
238 Hypatia	12.1		
240 Vanadis	11.7		
241 Germania	11.9		
245 Vera	11.5		
246 Asporina	12.4		
247 Eukrate	10.7		

Table 4.1 *(cont.)*

Asteroid	$M_{(B)}$	Date	Comments
250 Bettina	11.6		
253 Mathilde	12.6		
258 Tyche	11.1		
259 Aletheia	12.2		
261 Prymno	12.1		
264 Libussa	12.1		
266 Aline	11.8		
268 Adorca	12.5		
269 Justitia	12.9		
270 Anahita	10.8		
273 Atropos	12.8		
275 Sapientia	12.1		
278 Paulina	12.9		
283 Emma	12.7		
284 Amalia	11.8		
287 Nephthys	11.6		
304 Olga	11.8		
305 Gordonia	12.8		
306 Unitas	11.3		
308 Polyxo	12.3		
312 Pierretta	12.4		
313 Chaldaea	11.1		
317 Roxane	12.2		
322 Phaeo	11.9		
323 Brucia	11.1		
324 Bamberga	8.3		
325 Heidelberga	12.8		
326 Tamara	11.2		
335 Roberta	11.4		
336 Lacadiera	12.3		
337 Devosa	11.4		
344 Desiderata	9.7		
345 Tercidina	11.7		
346 Hermentaria	11.2		
347 Pariana	12.0		
349 Dembowska	10.3		
350 Ornamenta	12.5		
351 Yrsa	12.5		
352 Gisela	12.0		
354 Eleonora	10.3		

Table 4.1 *(cont.)*

Asteroid	$M_{(B)}$	Date	Comments
356 Liguria	11.0		
358 Apollonia	12.9		
359 Georgia	12.4		
360 Carlova	12.1		
361 Bononia	12.9		
362 Havnia	12.7		
363 Padua	13.0		
364 Isara	12.0		
365 Corduba	12.7		
369 Aëria	12.1		
371 Bohemia	12.8		
372 Palma	10.8		
374 Burgundia	13.0		
375 Ursula	11.8		
376 Geometria	11.8		
377 Campania	12.6		
379 Heunna	12.6		
380 Fiducia	13.0		
381 Myrrha	12.9		
382 Dodona	12.7		
384 Burdigala	12.9		
385 Ilmatar	11.3		
386 Siegena	10.9		
387 Aquitania	10.2		
389 Industria	11.9		
391 Ingeborg	12.2		
393 Lampetia	10.4		
394 Arduina	12.8		
397 Vienna	11.9		
402 Chloë	12.3		
404 Arsinoë	11.5		
405 Thia	10.6		
407 Arachne	12.5		
409 Aspasia	11.0		
410 Chloris	10.8		
413 Edburga	11.4		
415 Palatia	11.5		
416 Vaticana	10.9		
417 Suevia	12.9		
419 Aurelia	10.5		

Table 4.1 *(cont.)*

Asteroid	M$_{(B)}$	Date	Comments
422 Berolina	12.3		
423 Diotima	12.0		
425 Cornelia	12.6		
426 Hippo	12.5		
431 Nephele	12.9		
432 Pythia	11.7		
433 Eros	7.6		
434 Hungaria	12.7		
435 Ella	12.9		
437 Rhodia	12.3		
441 Bathilde	12.3		
442 Eichsfeldia	12.7		
444 Gyptis	11.1		
446 Aeternitas	12.7		
449 Hamburga	12.3		
451 Patientia	11.1		
454 Mathesis	12.6		
455 Bruchsalia	10.9		
458 Hercynia	13.0		
464 Megaira	12.8		
469 Argentina	13.0		
471 Papagena	9.8		
472 Roma	12.4		
475 Ocllo	12.2		
476 Hedwig	12.4		
477 Italia	12.6		
478 Tergeste	12.5		
479 Caprera	12.6		
480 Hansa	12.3		
481 Emita	12.1		
485 Genua	11.8		
487 Venetia	11.9		
488 Kreusa	11.8		
497 Iva	12.4		
498 Tokio	11.5		
500 Selinur	12.6		
503 Evelyn	12.3		
505 Cava	11.2		
506 Marion	13.0		
509 Iolanda	13.0		

Table 4.1 *(cont.)*

Asteroid	$M_{(B)}$	Date	Comments
510 Mabella	12.5		
511 Davida	10.2		
512 Taurinensis	12.2		
516 Amherstia	10.4		
519 Sylvania	12.4		
521 Brixia	10.6		
532 Herculina	9.0		
537 Pauly	12.4		
539 Pamina	13.0		
545 Messalina	12.8		
547 Praxedis	12.2		
550 Senta	12.1		
554 Peraga	11.3		
556 Phyllis	12.6		
563 Suleika	11.3		
564 Dudu	12.9		
578 Happelia	12.6		
579 Sidonia	12.4		
582 Olympia	11.9		
584 Semiramis	10.6		
593 Titania	12.1		
596 Scheila	12.5		
597 Bandusia	12.6		
598 Octavia	12.1		
599 Luisa	10.6		
602 Marianna	11.8		
606 Brangane	13.0		
618 Elfriede	12.9		
622 Esther	12.5		
626 Notburga	11.2		
631 Philippina	12.8		
639 Latona	12.5		
654 Zelinda	10.1		
660 Crescentia	12.5		
665 Sabine	12.9		
674 Rachele	11.0		
675 Ludmilla	11.3		
678 Fredegundis	12.2		
679 Pax	10.8		
680 Genoveva	12.4		

Table 4.1 *(cont.)*

Asteroid	$M_{(B)}$	Date	Comments
686 Gersuind	11.9		
690 Wratislavia	11.6		
694 Ekard	10.8		
695 Bella	11.6		
696 Leonora	12.4		
699 Hela	10.9		
702 Alauda	12.2		
704 Interamnia	10.1		
705 Erminia	12.8		
712 Boliviana	11.1		
714 Ulula	12.8		
726 Joella	13.0		
727 Nipponia	13.0		
735 Marghanna	11.5		
737 Arequipa	11.3		
739 Mandeville	12.1		
747 Winchester	10.0		
751 Faïna	11.6		
753 Tiflis	12.3		
758 Mancunia	12.4		
760 Massinga	12.2		
762 Pulcova	12.9		
772 Tanete	12.5		
774 Armor	12.9		
776 Berberica	11.2		
779 Nina	11.3		
783 Nora	12.8		
784 Pickeringia	12.9		
785 Zwetana	11.8		
786 Bredichina	12.9		
788 Hohensteina	12.8		
790 Pretoria	12.7		
796 Sarita	10.8		
804 Hispania	11.3		
814 Tauris	11.9		
849 Ara	12.1		
852 Wladilena	11.5		
886 Washingtonia	12.0		
887 Alinda	10.5		
888 Parysatis	12.6		

Table 4.1 *(cont.)*

Asteroid	$M_{(B)}$	Date	Comments
907 Rhoda	13.0		
914 Palisana	11.8		
925 Alphonsina	12.3		
931 Whittemora	12.7		
947 Monterosa	12.9		
952 Caia	12.4		
980 Anacostia	11.1		
984 Gretia	12.8		
1013 Tomecka	12.9		
1021 Flammario	11.3		
1036 Ganymed	7.8		
1048 Feodosia	12.8		
1093 Freda	12.1		
1116 Catriona	13.0		
1146 Biarmia	12.9		
1278 Kenya	12.7		
1310 Villigera	12.2		
1474 Beira	12.2		
1580 Betulia	10.6		
1627 Ivar	9.8		
1657 Roemera	12.5		
1659 Punkaharju	12.8		
1917 Cuyo	10.6		
1943 Anteros	11.1		
1980 Tezcatlipoca	9.9		
2000 Herschel	12.5		
2061 Anza	11.5		
2204 Lyyli	12.6		
2335 James	12.8		
2382 Nonie	13.0		
2608 Seneca	10.9		
3122 1981 ET3	7.2		
3199 Nefertiti	12.0		
3288 Seleucus	11.5		

4.5 Photography and electronic imaging

Seeing asteroids is a pleasant pastime, but a permanent record is not hard to make. Astrophotographic techniques ranging from the simplest beginner methods to some of the most difficult can be used to provide a challenge and satisfaction for everyone. The methods described here can be used by observers using photographic or electronic imaging techniques (see Section 3.6).

Imaging asteroids is no more difficult than imaging comets. In fact, while the techniques are the same, the simple stellar appearance of asteroids makes them an easier target overall, since no fine detail can be recorded.

At its simplest, this type of photography can be accomplished with only a 'normal' lens on a camera used wide open (a 50- or 55-mm $f/1.2$ to $f/2$ lens on a 35-mm camera), fast film (ISO 400 or faster), a cable release to open the shutter for a period of time, and a tripod to aim and hold the camera. The brightest asteroids can be recorded with this simple star-trail photography configuration using exposures of 10–20 seconds. [Fig 4.4]. A detailed star atlas will probably be necessary to identify the asteroid. A pair (or more) of exposures separated by hours, or a day, will show the motion of the asteroid against the background sky.

The addition of an equatorial mount and longer operating focal length will bring many more asteroids within a photographer's reach. Guiding on a nearby star, as in deep sky photography, allows many faint asteroids to be captured.

Figure 4.4   Vesta in Sagittarius. The bright 'star' near the center is actually the minor planet. Photograph by S. Edberg.

Interesting projects are possible with this equipment. The observer can obtain two or more images of the asteroid, maintaining the same guide star and star field. Then two photographs can be used as stereo pairs with the asteroid appearing to hover above or below the star field (Section 4.6).

With bright, fast-moving asteroids, skillful astrophotographers can face a different challenge. Maintain the same star field and guide star and, at regular time intervals, photograph the asteroid as it moves through the field. The result will be a multiple exposure showing the asteroid's motion across the frame. By using the same exposure time for each, it is possible to see if the asteroid's apparent magnitude changes during the observation period. (Note that the sky needs to be consistently clear throughout these exposures. Constant atmospheric transmission is needed for the duration of the sequence to show that any brightness variations are intrinsic to the asteroid.)

Fast-moving asteroids offer another possibility in leaving a trace when they are photographed. The image of the asteroid may be elongated, depending on the focal length used, exposure time, and the asteroid's apparent motion, when the telescope is guided at the sidereal rate (i.e. the telescope is exactly following the diurnal motion of the stars) [Figure 4.5]. The length of the trail, as men-

Figure 4.5   This pair of photographs shows the motion of Toutatis. A 10-minute exposure first captured the asteroid on one side of a star. One half hour later a 15-minute exposure shows the asteroid on the other side of the star. The star was occulted along a path passing several hundred kilometers north of this observing site. Photographs by S. Edberg.

101

tioned earlier, depends on the object's speed and the duration of the exposure. Earth-approaching asteroids may move as rapidly as a degree per day or much more.

The apparent motion of an asteroid makes it more difficult to record on film. Since the target is moving the photographic emulsion doesn't have the opportunity to build up an image as it does with the 'fixed' stars when the telescope is driven at the sidereal rate. To record rapidly moving or even slow, faint asteroids, the telescope should be driven at the target's computed rate of motion. This allows the asteroid's image to build up at one place on the emulsion so it can be seen, while the background stars will be streaks. Use the techniques described earlier for guiding on a comet (Section 3.6.2). Note that a separate guide telescope will be necessary.

4.6 Modern asteroid search methods

The discovery of new asteroids is a research area in which amateurs can participate. There are many successful amateur discoverers living in Japan.

To ferret out asteroids a 'blink' system or a stereomicroscope can be used. In a blink system, separate, magnifying optical systems are used to study the images, one for each image in the pair.

With a blink microscope or stereo blink comparator (Lazerson, 1984 and Mayer, 1988), the magnified view of each individual negative is sent to separate eyes, one at a time (thus, the blinking of the view). When the negatives are properly aligned, anything truly moving against the background stars in the field of view will appear to jump back and forth.

A pair of slide projectors may also be used, with a rotary shutter, to project alternating views of the pair of negatives (black stars on bright sky background) or positive slides (Mayer, 1977, 1978, 1984). Again, with proper alignment, moving objects jump back and forth while stars remain stationary. Electronic images can alternate on the computer screen with the same result as using slide projectors. Be aware that many people are uncomfortable watching a blinking, bright field when negatives are used in the search.

Another approach is to use a stereomicroscope. Around the turn of the century, Max Wolf of Heidelberg used a stereomicroscope to examine films taken with the astrograph there. But it was not until 1980 that Eugene Shoemaker established the use of the stereomicroscope as a valid tool in modern analysis of photographic film. His experience with using stereo viewing in geology led him to the idea that it could be a useful tool to record moving objects on photographic films of the sky.

The stereomicroscope consists of a binocular microscope adapted with prisms so that each film or plate projects through its own eyepiece. The observer looks through the binocular eyepiece so that the right eye sees the earlier film and the left eye sees the later film. In this way, objects moving retrograde appear to climb out of the film, and prograde objects appear below the level of the film. (An object moving so fast that it trails during the typical 6-minute exposure will appear as two trailed objects, not as a single one risen above the level of background stars.) Using a slow but continuous motion, Carolyn Shoemaker spends about 20 minutes to scan a pair of films covering a 9-degree-diameter circular field of sky.

Between 1980 and the end of 1991, the Shoemakers discovered about 750 interesting asteroids. Their first near-Earth asteroid (NEA) was an Amor, found in 1983 and later numbered and named 3199 Nefertiti after the queen consort of an Egyptian Pharaoh. They have also discovered 30 comets, as well as interesting non-solar system objects, like eclipsing or cataclysmic variable stars that change radically enough to be noticed in the 45-minute time interval between films.

4.7 Observing occultations by asteroids and of asteroids

Observations of an asteroid occulting a star can offer a real scientific payoff. Unlike that of a solar eclipse, the position of an asteroid's narrow occultation path has a notable margin of error. When positions of the minor planet and star are measured in the weeks and days before a scheduled occultation, the prediction of the occultation track can change by hundreds of kilometers.

Observing asteroidal occultations is a relatively new field which came of age, in a sense, with publicity surrounding the asteroid 433 Eros on 23 January 1975, when for two brief seconds the tiny planet could block the light of a star. The asteroid would be a mere 23 million km (14 million miles) from Earth on that night, and would occult Kappa Geminorum, a 3.6 magnitude star. Since Kappa Geminorum is so much farther away than Eros, the shadow of the asteroid crossing the Earth would be essentially the same size as Eros itself – only 22 km (13 miles). Radar observations conducted around the same time gave dimensions of 36 × 15 × 13 km (22 × 9 × 8 miles). This was David Levy's first attempt to observe such an occultation. At first it was assumed that the track of this occultation would sweep just west of Quebec City, Canada, but as time passed more accurate predictions of the racing asteroid indicated that the track would be further west, perhaps even over Montreal, where David Levy lived at the time [Fig. 4.6].

Thursday 23 January dawned bright and clear, but as the day wore on clouds gathered ominously. By nightfall it had begun to snow, and all over Montreal observers gave up in frustration as little Eros just passed us by. Successful observations of this event were made further east, along the eastern seaboard of the United States, just before the shadow headed out to sea.

In May 1983, asteroid 2 Pallas occulted the star 1 Vulpeculae in an event that was widely observed along its path across North America. A well-organized group spread along and across the occultation path in an attempt

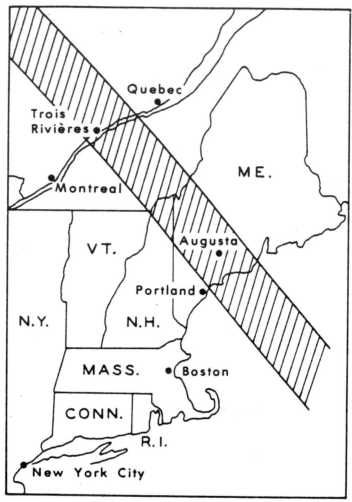

Figure 4.6   The predicted occultation path of Eros on 23 January 1975 as it crossed in front of Kappa Geminorum. The actual path shifted southwest, crossing through the southwest corner of Massachusetts and central Connecticut. Courtesy of Sky and Telescope.

to gather clues that could reveal the size and shape of the asteroid [Fig. 4.7]. Both authors of this book observed the star together for 20 minutes before and after the predicted time of the occultation event, as part of a network of observers stretching for many hundreds of kilometers along and across the path. Because we were north of the path, the asteroid did not dim the light of the star, but the negative observation was also important for it helped put an upper limit on the size of the asteroid. This experience was personally important for us, for it was our first meeting together and the start of our friendship.

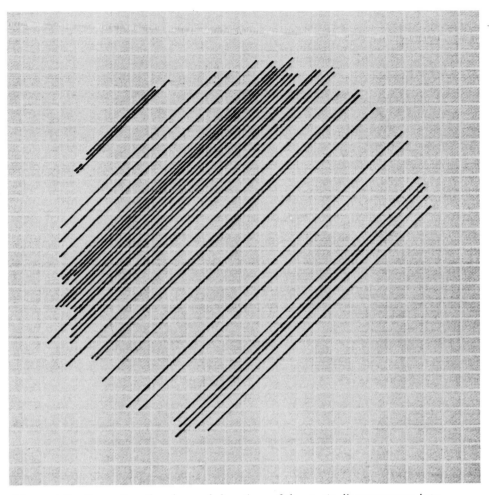

Figure 4.7 Converting the observed durations of the star's disappearance into distances by multiplying (duration) by (asteroid's velocity) gives the lengths of the chords across Pallas' cross-section as seen at the time of the event in May 1983. Projecting the geographic positions of observers relative to the asteroid's path on the Earth's surface provides the lateral spread. Combined, we see the asteroid's shape. Poor weather on the south side of the path prevented the full cross-section from being determined. A detailed report of results will be found in Dunham et al. (1990). Courtesy of D. Dunham and Sky and Telescope.

The best way to observe an occultation is through a fairly high-power eyepiece with a narrow field, so that you concentrate on the star that is about to be occulted. Familiarize yourself with the field around the star a day or two in advance so that you are absolutely certain that you have identified the star correctly. On the night of the event, begin observing at least 15 minutes before the predicted time, just in case the asteroid has a previously undiscovered companion that might also occult the star.

The work is exacting. Since an occultation typically lasts no more than a few seconds, if you turn away from the eyepiece for even an instant you could miss the event entirely. With the occultations by 4179 Toutatis, which made a close flyby of Earth in December 1992, event durations were frequently less than one second. Such events are more reliably timed photoelectrically.

You should have a radio that is playing government time signals and a tape recorder that records both the time signals and your comments. At the instant the occultation starts, say 'In!' so that the recorder will preserve your comment. When the star reappears, seconds later, call 'Out!' By later comparing your calls to the time signals on tape, you can reduce that time to the nearest tenth of a second, which is about as accurate as a visual observer can get.

Observing these occultations is an excellent project for astronomy clubs, since a thoroughly observed event requires coordination. The more observers at different sites along the length and width of the path, the better defined will be the path's location and projected size. This requires coordination and advance planning, as well as a well-motivated group of observers. A well-observed event can contribute to our understanding of the size and shape of an asteroid. Send your occultation report to the International Occultation Timing Association, a respected society headed for many years by David W. Dunham (see Appendix V for the address).

It is also possible to photographically record an asteroidal occultation. With very fast film loaded in the camera, aim a telescope + camera at the target star, with the clock drive OFF. Make a star trail photograph of the target star and other nearby stars. There will be a break in the trail of the target star if it is occulted. Careful measurement of the length of the break, when compared to the measured length of the complete star trail, will permit the duration of the occultation to be computed when it is scaled with the known duration of the exposure.

With the uncertainties inherent in occultation predictions it is sensible to have a field of view of at least one to two degrees along the long axis of the frame at the film plane. This translates to an effective focal length of 2000 mm to 1000 mm and a minimum exposure time (at the celestial equator) of eight to four minutes, respectively. Since the star is moving (as far as the film is concerned), a fast (low $f/$ ratio) telescope is desirable; evaluate the expression in Section 5.8.1 if necessary when selecting a telescope.

Asteroids are occasionally occulted by the Moon. These events are not common, especially for the brightest asteroids, but offer another method of determining the size of the object being hidden. The larger and closer asteroids will take a noticeable amount of time to disappear or reappear. From such timings, especially if they are made photoelectrically, we can determine the projected size of the asteroid facing the Earth at the time of the event.

Few events can inspire such excitement as the blocking of a star's light by a nearby rock in space. When you observe an occultation, you are watching the solar system in action, sharing one of its secrets with you.

4.8 Conclusion

For the study of asteroids, the future is very exciting indeed. More than many other scientific pursuits, studying asteroids is now seen as a subject with obvious social relevance, for it makes sense to know what might be supplying raw materials for our space settlements or blocking our way in the years and centuries ahead.

5

Meteors

5.1 Introduction

Meteors are centerpieces of the lives of children and lovers, who are entitled to wish upon them, and meteor observers, who are expected to brave frigid temperatures, mosquitoes, and the strains of fatigue to record them. The amateur meteor specialist enjoys the spectacle of the sky that greets his unaided vision, a sky unconfined by eyepiece or dome slot. Because meteor observing requires almost no equipment, and only the basic knowledge of the constellations and magnitudes – areas of astronomy appropriate for all beginning observers – one would think that meteors would be the most popular interest area for beginners.

Surprisingly, they are not. Meteor study is often ignored, even though other activities like deep sky observing and astrophotography are more demanding in equipment. Why is this so? Perhaps it is because the best conditions for meteor observing often occur after midnight, or maybe people just don't know how much fun meteor observing can be. Although meteors can be observed by one person alone, they are fun to watch as a team sport, conducted as groups assemble for 'shower parties' for the most important meteor streams.

Meteors are teachers of the sky, for observing meteors involves learning some basic facts about where the constellations are, the magnitude differences between stars, and the relationship of solar system objects to the Earth. Through observing meteors, one can get a clear illustration of the effects of perspective.

Amateur scientists are at a great advantage compared with their professional counterparts. We can gather our data at leisure. If we fail to produce, or feel the results are inferior, no one else need know about them. We do this work because we love it, not because we have to. We are poets as well, able to watch, enjoy, and learn from the falling meteors.

Few of the most talented amateur scientists, however, can resist the desire to share their experiences with others. Indeed, individual satisfaction with our own work is very much enhanced if we know the information will eventually be included as a basis for a scientific publication or otherwise used by a professional scientist. In the past, amateur and professional astronomers have enjoyed a good working relationship in meteor science as well as in other fields.

In the decades since the 1957–58 International Geophysical Year (IGY), when professional–amateur cooperation set a new standard in meteor work, professional meteor science in North America has dwindled. But there are some encouraging signs that this may change in the years to come. Until this happens, it is up to amateurs to carry on the observing traditions of the past so that their work can have maximum scientific impact.

5.2 Historical notes

While meteor showers have been chronicled for centuries, the interest of serious observers in these events intensified in 1833 with the Leonid meteor storm. In the early 1920s Charles P. Olivier founded the American Meteor Society (AMS), a group of dedicated meteor observers whose observations would build up a central archival file of meteor work done over many years. Dr. Olivier endorsed a mode of operation that concentrated on the efforts of single observers and not with group reports. This operation was highly successful, and when Dr. David Meisel succeeded Olivier as Director, the AMS continued its archival base of single observer statistics. Professionals realized the considerable value of amateur contributions in the 1950s when planning for the International Geophysical Year. Large numbers of observations were accumulated. This international program of cooperation between countries included professional and amateur astronomers alike.

In no country was this program taken more seriously than Canada, where Dr. Peter Millman of the National Research Council carefully prepared an observing form that would be a model of simplicity and could be understood by almost anyone. For each meteor, only two pieces of data were needed: magnitude and shower membership. The time column was left unlined, so that observers could record time to the hour, minute, or second. Here was an 'observer-friendly' program that, if successful, could attract hundreds of Canadian amateur astronomers.

It did. From the time the program began at the dawn of the IGY in 1957 to its official close almost fifteen years later, groups of observers across the country organized major observing projects for most of the major annual showers. The observing became so intense that for some astronomy societies in Canada, meteor observing was the major observational activity. The program generated such interest that even today, years after its official close, amateurs still use the forms and the basic procedures to observe meteors for themselves. Unlike the AMS work, the IGY program encouraged group observing.

From years of observing by the AMS, IGY, and other groups, amateurs have achieved a monumental file of archival data that shows the changing strength of meteor showers over three-fourths of a century. These data are, in their own way, as valuable as those obtained by the most modern radio and radar scans, simply because they stretch for so long. The data required are easily obtained and provide both the novice and the experienced observer with an opportunity to contribute to an area in which few professional astronomers participate.

5.3 Meteor astronomy

A meteor's existence usually begins in a comet, or less often in an asteroid. A dust particle whose density is about 0.3 g/cm³ (about the same as some foamed plastics) is released and blown away when cometary ices in the nucleus sublimate in the heat of the Sun. It becomes a part of the comet's dust tail. Eventually the tail disperses but the particle continues circling the Sun while its orbit is continually changed, or perturbed, by various effects due to the planets or the Sun itself. Eventually its orbit may intersect with Earth's orbit, perhaps millennia after it was released by the comet, and if the two find themselves at the same place at the same time this meteoroid enters the Earth's atmosphere. Its velocity will be between 11 km/s (7.5 miles/s) and 72 km/s (45 miles/s), depending on whether the collision occurs with one body overrunning the other or if it is head-on, or most likely somewhere in between.

Observers viewing the collision will typically see a fast-moving streak of light in the night sky – a 'falling star' [Fig. 5.1]. Some meteors exhibit a wake

Figure 5.1   Two Perseid meteors including one with a modest flare captured on 12 August 1980. Photograph by S. Edberg.

or trail, which is a short-term streak along the path followed by the meteor. The brightest meteors may leave a persistent train, which is visible for seconds or even minutes before dispersing. The meteor may first be visible as high as 120 km (75 miles) and if it is bright it may not burn out until it is as low as 60 km (35 miles). Usually meteors are first seen at heights of 80 to 100 km (50 to 60 miles) and disappear at about 50 km (30 miles).

A typical meteor is an event in which a sand-grain-size to pea-size particle called a meteoroid is slowed by atmospheric friction, and in the process exciting its atoms as it disintegrates and ionizing the air around it to create the moving point of light and wake we see. Smaller particles, which form the 'bulk' of cometary dust tails, are gently slowed by the atmosphere without heating and waft their way down as micrometeorites, perhaps to form the nuclei of rain drops. About one-third of the dust recovered in the stratosphere appears to have an origin with the asteroids. Micrometeorites scatter the sunlight seen as the zodiacal light, in its manifestations as the zodiacal light pyramid, the zodiacal band, and the gegenschein (Chapter 6).

Larger meteoroids are seen as fireballs, some brighter than Venus or even the Moon [Fig. 5.2]. A fireball seen to explode (since normal meteors rarely do) is called a bolide. After a fireball fades away any remaining material will fall to Earth as a meteorite, and especially if detonations are heard, (even minutes after the fireball) a search for the meteorite is in order. To recover a

Figure 5.2   A Perseid fireball that flared to magnitude -8 on 12 August 1978 at 1109 UT. Photograph by S. Edberg.

111

meteorite, accurate observations of the altitude and azimuth of the beginning and end points of the fireball are necessary from many sites. Triangulation is then used to determine the air path and project the probable point of impact.

Recent research, using known asteroid orbits (Drummond, 1991) or using computed fireball orbits (Halliday, Blackwell, and Griffin, 1990), suggests that there are asteroid streams. Members of these streams have closely related Amor-type orbits (Section 4.3.2). Such streams could lead to long-duration, low-rate showers of very slow-moving sunset meteors with long durations. The showers would not have a well-defined peak.

Cometary meteoroids are generally fluffy and burn up in the Earth's atmosphere before reaching the ground. Meteoroids believed to originate in the asteroid belt between Jupiter and Mars have been recovered as stony, stony-iron, or iron meteorites.

As one might imagine, if one cometary dust particle can eventually collide with Earth, many can. Meteor streams contain meteoroids traveling in roughly parallel orbits, and the collision of many particles with Earth's atmosphere causes a meteor shower. There are a few major annual showers (tens of meteors per hour), and a large number of minor ones (a handful of meteors per hour). On rare occasions, large numbers (thousands) of meteors seen typically in a short (hours long) period of time are referred to as a meteor storm. Even in a meteor storm, the meteoroids are kilometers apart in space. Meteor storms reflect the clumpiness of meteoroid distribution in the orbit of a comet, while the annual showers reflect the even spread of particles along an orbit which the Earth intersects at least once a year (and no more than twice a year). It is interesting to note that were the Earth to pass through a comet trail, thousands of meteors per *second* might be seen!

Because meteoroids in a stream travel in parallel paths, a meteor shower appears to radiate from a small area in the sky called the radiant. The apparent intersection of the paths of meteors (that are really traveling along parallel paths) at a point in the sky is an optical illusion, much like the apparent intersection of parallel railroad tracks at the distant horizon: it is a matter of perspective.

A meteor shower radiant [Fig. 5.3]. is not a point in the sky, but rather a small area. The size of the area indicates the age, or equivalently, the dispersion of meteoroids in a stream. A small radiant size indicates a youthful, hardly dispersed stream. A few meteor showers exhibit several radiants a few degrees apart in the sky. Meteor showers are usually named after the constellation in which the radiant resides during the shower's peak. Because the Earth moves relative to the meteoroid orbits, the apparent radiant shifts from night to night.

Many meteors appear to radiate from areas other than those of the main showers. While they may be members of a minor shower, they may also be sporadic meteors, apparent loners in their atmospheric swan song. Sporadic meteors are thought to be the remains of ancient streams which have had

Figure 5.3   Three Perseids, whose paths were parallel in flight, are traced back to the shower radiant. The divergence from an apparent radiant is an effect of perspective, just as the apparent meeting of railroad tracks in the distance is. Their trails are short because they were captured close to the radiant.

their orbits dispersed by planetary gravitational perturbations and the pressure of sunlight.

Peak hourly meteor rates are usually seen when the Earth crosses the average orbital plane of the shower meteoroids. Youthful showers are narrowly peaked in meteor rate, with the period when the shower is one quarter or more of its peak strength lasting as little as a few hours. Older showers may be many days 'wide', because the meteoroids' orbits have changed with time.

More meteors are seen in the hours before dawn than after sunset. This effect occurs because in the morning hours meteoroids are colliding more nearly head-on, rather than catching up with Earth as they do in the evening hours. The closer to the zenith the radiant is, the higher the meteor rate usually will be. For a large number of showers the radiants are highest just before dawn.

There is an unexplained seasonal variation as well. Overall, more meteors are seen over the July–December period than are seen over the January–June time frame.

113

5.4 Meteor showers

From a dark country sky, one can observe lots of meteors any night of the year. We know of one stalwart person observing alone, who counted some 200 meteors on a night with no major showers.

Of course, nights of major showers will be more interesting, and should offer opportunities to compare the strength, duration, and radiants of the different meteor streams through which our planet passes. On other nights, there are often one or more minor showers for the novice to 'discover' and appreciate. Here are short descriptions of the more reliable major showers. The rates quoted are called zenithal hourly rates (ZHRs), and assume the ideal circumstance of a single observer observing under a dark, moonless sky, limiting magnitude 6.5, while the radiant is at the zenith. The references mentioned can be consulted for further details.

5.4.1 Major annual showers

Quadrantids

This January shower offers up to 200 meteors per hour (190 in 1965) per observer at maximum, competition for the August Perseids. Some amateurs believe that the reason these meteors are not better observed is that the January nights are so cold, but this is perhaps only part of the answer. Although the Quadrantids are strong, they last only for a few hours on the night of maximum. Unless the Moon is out of the night sky, and unless the maximum occurs at night and not during daylight, this shower is unspectacular. Even when the circumstances are perfect, this shower (and others) can be poor if the Earth is not passing through a dense portion of the stream. The 1973 and 1985 displays occurred under ideal conditions and were superb. The name is based on an ancient and now abandoned constellation called Quadrans Muralis – the mural quadrant, surrounded by Ursa Major, Boötes, Hercules, and Draco.

Lyrids

We have to wait almost four months before the next major shower occurs, the Lyrids. These meteors appear at the rate of about 15 per hour per observer on the night of maximum, 20/21 April. However, in 1803 observers in eastern North America observed some 700 meteors in one hour, and activity reaching about 80 per hour occurred briefly in 1982, a dramatic increase characterized by fainter than usual meteors. The shower originates from Comet Thatcher 1861 I.

Eta Aquarids

This is one of two annual showers originating from Halley's Comet. Related meteors are identifiable from 21 April to 25 May, with a broad maximum around 3 May. These meteors appear as fast streaks. The brightest of the Eta Aquarids leave long-lasting trains. Because they are on the outbound leg of their orbits, these meteors arrive mainly in daylight; thus the night time dark sky observation interval is short and occurs just before dawn.

Delta Aquarids

A delight to observe, this shower's midsummer (late July) maximum means that it is ideal for children on summer vacation; this could be the shower that turns them into young astronomers. The meteors are faint, of medium speed (considerably slower than the neighboring Perseids) and offer a single-observer hourly rate of 20 for observers at lower latitudes. These meteors tend to vaporize over long paths. Hamid and Whipple (1963) suggested that the Quadrantid meteors were once part of the Delta Aquarid stream, their perihelion distances and orbital inclinations having been similar until about AD 700.

Perseids

This August shower can be a magnificent spectacle. We have seen Perseids coming in such rapid succession that counting and recording were difficult, followed by slack periods with little activity. We have also seen 'firework' Perseids, when two or more meteors appear almost simultaneously and fly off in different directions. Shadow-casting fireballs are not uncommon. In a somewhat limited and theoretical sense, this is a period that should be avoided by all but the most experienced observers since several showers are active at the same time – Southern Delta Aquarids, Northern Delta Aquarids, Alpha Capricornids, Southern Iota Aquarids, Northern Iota Aquarids, Kappa Cygnids, Upsilon Pegasids, and Alpha Ursa Majorids. With all this activity it is sometimes difficult to tell from which shower a meteor originated. However, it is obvious that these showers should be enjoyed by as many people as we can possibly gather. Their differing behavior from other showers of this time (they are faster than the Delta Aquarids and they sometimes fragment) will offer even beginning meteor observers a real education in how meteors act. The 1991 shower peaked at 300–400 meteors per hour for fortunate visual observers in eastern Asia and radio observers making measurements in daylight elsewhere (MacRobert, 1992). Storm conditions prevailed in 1992 for privileged observers in the eastern hemisphere, though the observed rates were only in the hundreds per hour due to the gibbous Moon. These observations presaged the recovery of the parent comet, P/Swift-Tuttle, in September of that year. The 1993 Perseids were looked upon with considerable anticipation (Rao, 1993) and NASA decided to postpone the launch of the space shuttle Discovery to avoid any chance of collison damage. Ultimately, reports of hundreds per hour were received from observers situated in the longitude zone of the Mediterranean Sea and impressive activity was reported from North America (IAU *Circulars* 5841 and 5843). Cosmonauts aboard the Russian space station Mir reported taking many small hits, which were audible (D. Talent, private communication).

Orionids

Often as many as 25 meteors per hour are seen in this October shower. They are easily identified not only from their several radiants near Orion but also from their speed: at 66 km/s, they appear as fast streaks, faster by a hair than their sisters, the Eta Aquarids. This shower is a meteoric offspring of Comet Halley. And like the Eta Aquarids, the brightest of their family tend to leave long-lasting trains. Fireballs are somewhat more common about 3 days after maximum.

Taurids

Consisting of several related streams, this shower's maximum rate is 15 per hour, peaking between 1 November and 3 November. At 28 km/s, the slowest of the major shower meteors, they average a little brighter than the other major showers of this time, except the Leonids. Many fireballs reported in the months of October, November, and December belong to this stream. The streams originate from Periodic Comet Encke but in a special way; in 1950 Whipple and Hamid proposed that they were produced suddenly as material was discharged from the comet about 4700 years ago. About 1500 years ago, they add, a second ejection of material added to the streams. However, the parent of this new material was now a different comet with greater aphelion distance that had earlier separated from Encke. More recently, Olsson-Steel (1988) has used meteor radar orbit surveys to show that Oljato, 1984 KB, 1982 TA, and 5025 P-L, as well as P/Encke, are all closely associated with the Taurids.

Leonids

This moderately active shower in November has a normal maximum rate of 15 per hour for an observer working alone. But every 33 years, the orbital period of their parent comet P/Tempel–Tuttle (1965 IV and 1866 I), the Earth crosses the densest concentration of meteors and a storm often results during which the maximum is measured not in meteors per hour but in meteors per second. In 1966 some observers estimated 40 per second. The years preceding and following a storm year show enhanced activity, with Stephen Edberg noting 90 per hour for one hour in 1968. The 1969 shower was notable not so much for its numbers as for the numerous ruby red meteors Edberg and fellow observers noted. At 71 km/s, these are the fastest major shower meteors. They also offer the consistently brightest meteors of any annual shower, and frequently leave persistent trains. Even with low rates, Leonid maxima are interesting to watch. The 1990s offer the possibility of increasing rates, as occurred in the 1960s through 1966. Culmination of the potential increase, perhaps to storm level, will come with the 1998 and 1999 showers (Yeomans, 1981b, and Croswell, 1991). An International Leonid Meteor Watch is being organized by the International Meteor Organization (Appendix V). Many experienced observers scattered around the world are sought for this cooperative effort. There is plenty of time to gain the necessary experience if you start observing meteors now.

Geminids

This is one of the best showers of the year, and occurs in December. The Geminids offer a maximum of as much as 75 per hour, and stretch out for 2.5 days on either side of maximum. The night before and the night of maximum often are the best nights of the year to observe large numbers of meteors. Their relatively slow speeds add to the fun of watching this magnificent meteor shower, and to the beauty of their brightest members as they arc gracefully across an expanse of sky. Using data from the IRAS satellite in 1983, Simon Green and John Davies discovered asteroid 1983 TB, now called 3200 Phaethon. Although this object displays no cometary activity, its orbital match with the meteors suggests its parenthood of the Geminid stream. The Geminid meteoroids are higher density than those in other streams. This may be due to their close perihelia (0.14 AU). Such intense warming might make them more like cometary surface crusts rather than internal cometary material.

Ursids

Lasting from 17 to 24 December, this shower peaks on the 22nd. Discovered by the famous British observer William Denning, this shower typically has rates between 10 and 15 meteors per hour, and often the ZHR is as low as 2. However in 1945 A. Becvár (who prepared the *Skalnate–Pleso Atlas of the Heavens*) observed more than 100 in an hour, and the ZHR was more than 25 in 1979. Although the meteors tend to be faint, observers have observed meteors of magnitude 0 or brighter. The shower is probably associated with Periodic Comet Tuttle, first seen by Méchain in 1790 but named for Tuttle who rediscovered it in 1858.

5.4.2 List of annual meteor showers

Our list of meteor showers in Table 5.1 is designed to give you an idea of meteor activity at a glance and suggests the strength of most of the major showers. Adapted from two separate lists in Dr. Peter Millman's tables in the *Observer's Handbook of the Royal Astronomical Society of Canada*, the list summar-

Table 5.1. *Annual meteor showers*

Shower	Date	Rate	Speed	Duration
January				
*Quadrantids	3	40	41	1.1 days
February				
Delta Leonids	26		23	5 Feb.–19 Mar.
April				
Sigma Leonids	17		20	21 Mar.–13 May
*Lyrids	22	15	48	2 days
May				
*Eta Aquarids	5	20	65	3 days
June				
Tau Herculids	3		15	19 May–14 Jun.
July				
*S. Delta Aquarids	30	20	41	7 days
Alpha Capricornids	30	5	23	15 Jul.–10 Aug.
August				
S. Iota Aquarids	5	1–5	34	15 Jul.–25 Aug.
N. Delta Aquarids	12	2–5	42	15 Jul.–25 Aug.
*Perseids	12	40–100	60	4.6 days
Kappa Cygnids	18		25	9 Aug.–6 Oct.
N. Iota Aquarids	20	1–2	31	15 Jul.–20 Sep.
September				
S. Piscids	20		26	31 Aug.–2 Nov.
October				
Annual Andromedids	3		18–23	25 Sep.–12 Nov.
N. Piscids	12		29	25 Sep.–19 Oct.
*Orionids	21	25	66	2 days
November				
*S. Taurids	3 ⎱	15	28	—
N. Taurids	13 ⎰		29	19 Sep.–1 Dec.
*Leonids	18	15	71	—
December				
*Geminids	14	75	35	2.6 days
*Ursids	22	5	34	2 days
Coma Berenicids			65	12 Dec.–23 Jan.

izes what you may expect during the year. Leap years and other considerations could cause small changes in the dates of maxima.

For those of you who wish more detailed information on radiant positions and who want specifics on rarer showers, we also present a second list in Appendix III. We also recommend Gary Kronk's *Meteor Showers: A Descriptive Catalog*. Published in 1988, it describes more than 80 showers from historical, orbital, and observational viewpoints.

5.5 The meteor observer

5.5.1 Planning: moonlight and weather

Observing sessions are usually more enjoyable if you plan them beforehand. If the shower you are thinking about observing has its peak a week before full Moon, for instance, your observing circumstances would be far better in the moonless hours before dawn instead of in the evening.

Learn your local weather patterns. Planning to observe any shower between November and January from the northeastern US, or during July and August in the southwest, is a risky business since the weather is so uncertain. We remember Novembers with a total of one or two clear nights, and organizing a meteor watch during that time can be a depressing exercise. However, remember that you definitely won't see any meteors if you don't at least try. Plan some observations, and don't be too disappointed if the weather does not cooperate.

5.5.2 Comfort

As in most branches of astronomical observing, meteor watching is more effective if you are comfortable. If you stand up during your session, straining

Notes for Table 5.1:
A * indicates that the shower is a major visual one. In the 'Duration' column, the digit indicates the normal duration, in days, to approximately $\frac{1}{4}$ the maximum strength.
Minor showers often represent older streams without strong maxima. Thus, the dates presented in the 'Duration' column indicate when meteors from the shower might be visible.
The 'Date' column lists the average annual date of maximum.
The 'Rate' column refers to Zenithal Hourly Rate (ZHR) for major showers and indicates the number of meteors per hour per single observer that could be expected on an 'average' night of maximum with the radiant at the zenith. (Since these requirements are rarely met the observed rate may be much less.) Rates left blank are typically 0–3 per night.
The 'Speed' column lists the typical meteoric velocity relative to the Earth in km/s.
Source: We thank Peter Millman, whose meteor shower list in the RASC *Observer's Handbook* formed the basis for these data, and Norman W. McLeod III, whose insightful comments added to the material.

your neck as you look skyward for meteors, your total will be lower. The best possible way to observe meteors is from a reclining position, on a lawn chair with an adjustable back. Set up so that your eyes naturally fall on a position 45–50 degrees above the horizon. The back can be raised if you wish to observe the part of the sky below 45 degrees, or lowered if you wish to watch nearer the zenith. These chairs are superior to lying on the ground as they protect you somewhat from the dirt and bugs there.

Two Canadian groups who went into meteor observing in a serious way during the IGY and the decade after designed a set of eight enclosures, euphemistically called 'coffins', into which the observers could crawl [Fig. 5.4]. These wooden structures were completely enclosed, except for a small space for the head. The observers were thus kept warm in the cold nights that Canadians must put up with throughout most of the year, and also the structures helped somewhat to keep mosquitoes out during summer.

The 'coffins' were used by the National Research Council team, and also by the Meteor section of the Royal Astronomical Society of Canada's Ottawa Centre. These were serious observers: Kenneth Hewitt-White once observed Geminids this way for an entire night from 4:30 p.m. to 7:00 a.m. the following morning. But Ken thought it worth the effort: 'It pushed our total for the year to 3000 meteors!' In 1992 Rolf Meier redesigned the enclosures so that they could be taken apart for transport.

Figure 5.4   These 'coffins' keep observers like Matthew Meier warm during Canada's long, cold winter nights and provide some protection from mosquitoes during other times of year. Photograph by R. and L. Meier.

Clothing is also important. Some observers layer their clothing, starting with a light jacket or sweater, and gradually adding on more layers as the temperature falls. During summer you should be careful about insect repellant, making sure that you have enough of it to protect you from mosquitoes. On the other hand, insect repellent reacts chemically with plastic items, like forks, chairs, and tape recording equipment.

5.5.3 Staying awake

That pleasant, relaxed feeling that you get when you are lying in bed is an important requirement: spend some time making sure you will enjoy your session. We don't write this facetiously; your eyes will actually see more if your mind doesn't have to worry about stiffness in your back or neck.

Getting too comfortable, of course, increases the risk of actually falling asleep, and it is the considered opinion of both authors that this will decrease the number of meteors you see. We could say that the thrill of meteor watching should be sufficient to keep you awake, but this may be too much to expect on nights when the counts are low. Totals will drop when people tire, and drifting off from time to time is a possibility even for experienced people. A nap may help the observer stay awake with rested eyes.

Music: Even though music adds noise and could interfere with a central recorder hearing what you report, it can provide a level of comfort and interest that will help keep you awake during a long night. If you are alone, it is especially helpful to have some background music to help pass what could turn into a long night.

Conversation: If you are in a team, keeping up a conversation, possibly about items unrelated to meteors, can help keep observers alert. Of course, each time someone sees a meteor and calls 'TIME!' the conversation stops. The best way to keep alert is through intelligent scheduling. Especially if you are tired, don't plan to observe for more than an hour at a time, and follow each observing stretch with at least a half hour period for rest or coffee.

5.5.4 Safety

Alone or with a group, the observer should leave word as to destination, estimated time of arrival, route, and planned time of return. Citizens' Band (CB) or other amateur radio bands can be used to maintain contact with observers on the road or at the observing site. Normal supplies for outdoor visits, like insect repellant, first aid materials, snacks, water, toilet paper, and other items should be included in the observing kit.

5.6 Observing overview

Meteor observing is both simple and complex, challenging and fun. There are several distinct types of meteor observing, from simple visual counts to rather complex photographic or radio observations. Let us briefly explore some of these types.

Visual counts and radiant determination

This is the most popular type of meteor observing, and little wonder – visual observing is the traditional way of collecting useful data on meteor streams: their intensity, their members' brightness, and their duration. Although nights of major showers attract many observers, observers who record meteors every clear night have the best chance of discovering new streams.

You can observe meteors alone or in a group. If you watch by yourself, do not try to scan the entire sky. Choose one quadrant and concentrate on it. And be careful in your choice of quadrant: the one containing the radiant may not produce the most meteors. Choose an area about 60 degrees from the radiant, and of course in the darkest part of the sky.

Telescopic meteors

It has been theorized, and reported, that some showers emphasize faint meteors that are not visible with the unaided eye, and that often the number of faint telescopic meteors does not increase during times of some major showers. Telescopes with short focal ratios ($f/4$ or $f/5$) and low powers are best for this type of observation. Even annual counts of telescopic meteors seen in the course of other astronomical observations have value.

Fireballs

It is important to record accurately any fireball you may be lucky enough to see. A fireball is defined as a meteor with a visual magnitude of -4, approximately the brightness of Venus, or brighter. Fireballs brighter than -8 may have produced meteorites and are therefore of considerable interest.

Meteor photography

This can be a most useful practice. For starters, it is easiest to use your camera's ordinary lens of 50 mm focal length (or so), pointed 45–60 degrees from the radiant. Or, try pointing one directly at the radiant. Remember that depending on the speed of your film and the condition of the sky, you probably won't record any meteor fainter than magnitude $+1$.

Radio Observation

The simplest methods are suitable for beginners and are interesting and fun. Tune an FM radio to a distant station that cannot ordinarily be heard, and then listen for enhancements a few seconds long in the signal. Radio observing can also be done very elaborately, with a knowledge of amateur radio techniques and computers necessary.

5.7 Visual observing procedures

Meteor shower observing is simple and easy, and yet it requires patience, care, and perseverance. For each meteor you should record the time, magnitude, shower membership, and comments on anything unusual. Detailed observations do, however, require a great deal of practice to decrease a psychological bias that is inherent in visual work. Regular viewing of as many meteor showers as possible, throughout the year, will help develop and maintain a high degree of skill for these observations.

5.7.1 Weather report

This important record of your observations should be noted every half hour or more often if conditions change rapidly. The various patterns that can affect your totals and the accuracy of your counts are listed below. Recording the limiting naked eye magnitude will assist the interpretation of your meteor report.

The limiting magnitude can be estimated using star charts marked with stellar magnitudes (e.g., *The AAVSO Variable Star Atlas* (Scovil, 1990)) by finding the faintest star you *can* see and the brightest star you *cannot* see: the limit is between them. Another method is to count the number of stars visible in a specified area. Roggemans (1987) contains twenty such areas and their corresponding star lists.

(1) *Cumulus and stratus clouds*: These clouds are often thick and opaque and prevent you from seeing any meteors in the portions of the sky they cover. They differ from:

(2) *Cirrus clouds*: Thin cirrus is especially annoying because it can affect your counts without your knowing it. In a dark, moonless, country sky, cirrus might be difficult to detect. The sky might have a 'soupy' appearance, and the patterns of the constellations might look different. Bright stars might have faint haloes or glows, and the fainter stars might not be visible at all. Make a note if haze or stratospheric volcanic dust are affecting your observations as well.

(3) *Moonlight*: Since moonlight obviously affects the number of meteors you may see, it is important not only to record the time of moonrise but also to plan your sessions around it. A Perseid watch held around first quarter Moon on 6 August would be more easily analyzed than one conducted on the night of maximum a week later with a full

Moon. In the Time Record section of the IGY report form or in the notes on any other report you compile give the Moon's age, or at least a general indication of the lunar phase, including waxing or waning and new, crescent, quarter, gibbous or full.

(4) *Lightning*: Lightning from distant thunderstorms could be misinterpreted as faint meteor counts. Note its presence in your weather report.

(5) *Wind*: If the wind is uncomfortably strong, it could adversely affect the quality of the observations you make and should therefore be recorded.

(6) *Fatigue*: Even though degree of lassitude is subjective, it should be noted in your record as it can affect the accuracy of your work. If sleepiness is seriously influencing the accuracy of your counts, you may want to interrupt your observations.

5.7.2 Time

Whether you use a tape recorder or a central 'person recorder' you should announce the appearance of a meteor with a one-word alert, like 'TIME!' But how accurately the appearance of a meteor should be timed is a matter of hot debate, and the range of desired accuracy is large. When the International Halley Watch was collecting meteor shower reports, its required precision was only one hour for hourly count statistics; they simply wanted to know the number of meteors seen in an hour of observing. Other groups want each meteor recorded to the nearest minute or second. Perhaps Canada's Visual Meteor Program of the International Geophysical Year offered a fair compromise, asking that you record times no less frequently than every ten minutes, or five during heavy showers. If you use a tape recorder, there is no reason you cannot have each meteor recorded to the nearest second. Simply have radio time signals recording into the machine as well as your meteor reports.

Always record fireballs or other unusual meteors to the nearest second. It is always possible that a fireball's progenitor might land somewhere. In that circumstance an accurate description of the path, including the time of its appearance, is essential to locating the downed meteorite and perhaps reconstructing its solar orbit (Section 5.7.10). Other unusual meteors – long lasting ones that could be satellites reentering – should be timed accurately so other people at another site are able to confirm the event.

In the past, there has been some debate between meteor observers over when to begin hourly counts of meteors being recorded with a chronological precision of one hour. The three choices are: anytime, on the UT hour (or half hour), or corrected to the standard meridian.

'Anytime' is fine for casual observations not intended for comparison with others. Starting on the UT hour is one standard that permits easy intercomparison of hourly counts by observers anywhere. It makes it easier to study the physical correlation of meteor rates with the Earth's position in its orbit and also the meteoroid's orbit.

Correcting to the standard meridian means starting an hourly count on the hour, corrected by four minutes per degree of longitude that you are from the standard time zone meridian. Standard meridians are set every 15 degrees from the 0 degree meridian at Greenwich, England. Thus, if you are east of your meridian by 2 degrees, you would begin your observation period 8 minutes before the hour. This method of time keeping permits meteor counts from various sites to be compared directly with respect to the angular altitude of the radiant.

5.7.3 Magnitude

Estimating the brightness of a meteor should be a straightforward operation, but some observers have difficulty with this operation. There are three basic problems: (1) some people tend to overestimate, (2) others tend to underestimate, and (3) still others have an inconsistent pattern. When your eye is attracted by the sudden appearance of a meteor, the sense of surprise you feel can lead to overestimating. So long as you are aware of this pitfall, you should try to avoid it. Underestimating magnitudes is, in our opinion, not as common a problem, but it has been said to arise when the observer sees a meteor only after it has completed most of its appearance.

An observer can also be inconsistent, with some estimates much brighter than they should be, and others fainter. Such a difficulty arises when the observer is inexperienced or does not understand the magnitude scale. If that is the case, the problem is easily corrected by a short course in relative star brightnesses. Another cause might be a difficulty with perception, something that can be overcome with practice in estimating magnitudes.

It is a good idea to acquaint the observers with the standard stars to which the meteors will be compared a night or two before observing begins. If a magnitude estimate is uncertain, because of passing clouds, inattention, or any other reason, the uncertainty should be noted in the record by a '?' after the recorded magnitude.

We suggest using the following 'standard' magnitude stars for estimating the brightness of a meteor:

Brighter than 0	
Sun	-26.5
Full Moon	-12 to -14
Venus	-4.6 (maximum brightness)
Jupiter	-2.5 (maximum brightness)
Sirius	-1.5
Canopus	-1

<u>0</u>	
Arcturus	−0.1
Vega	0.0
Capella	+0.1
<u>0.5</u>	
Achernar	0.5
Procyon	0.5
<u>1</u>	
Spica	1.0
Altair	1.0
Deneb	1.3
Aldebaran	1.0
Pollux	1.0
<u>1.5</u>	
Regulus	1.4
<u>2</u>	
Alpha Persei	2.0
Beta Aurigae	2.0
Alpha Andromedae	2.1
Alpha Arietis	2.0
Alpha Ursae Majoris	2.0
Polaris	2.0
Beta Ursae Minoris	2.0
Gamma Geminorum	1.9
Gamma Leonis	2.0
Alpha Ophiuchi	2.0
<u>2.5</u>	
Delta Leonis	2.5
Gamma Ursae Majoris	2.4
Epsilon Cygni	2.5
Alpha Cephei	2.4
Alpha Pegasi	2.5
<u>3</u>	
Beta Trianguli	3.0
Epsilon Geminorum	3.0
Gamma Boötis	3.1
Gamma Ursae Minoris	3.0
Alpha Aquarii	2.9
Eta Pegasi	3.0
<u>3.5</u>	
Eta Ceti	3.4
Beta Boötis	3.4
Alpha Trianguli	3.4

Epsilon Tauri	3.5
Lambda Aquilae	3.4

5.7.4 Color

Many new observers do not realize that the brighter meteors may display some color. Be alert for this effect, since a pronounced color can give a hint of the meteor's composition. Stephen Edberg and fellow observers were privileged to note numerous ruby-red Leonids in 1969. Fireballs are often reported as pastel green or blue, sometimes accompanied by orange sparks.

5.7.5 Shower membership

If you have a basic understanding of perspective, you should have no trouble determining the shower to which a meteor belongs. Tracing an imaginary path back to a likely radiant is an easy process which becomes second nature after a little practice. The summer showers offer more of a challenge, since two major radiants and several minor ones may be active at the same time. It is possible to have difficulty deciding to which of these showers a meteor belongs, especially if it streaks toward the southwest. But even then, remember that the shower meteors have particular characteristics, such as speed, that can assist in deciding their shower membership. If in real doubt assign the meteor to a non-shower category. Do not assign any meteor to membership in a minor shower unless you are experienced enough to be absolutely certain. Also, keep in mind that sporadics can come from any direction, including that of a shower radiant. You should suspect this if the uncertain meteor is travelling at a speed greatly different from what you expect from the shower.

5.7.6 Comments

Qualitative impressions or quantitative data on each shower can be helpful. These could take the form of comments like 'large number of fragmenting meteors' or 'about 20 percent of observed meteors had a reddish tinge'.

Was a particular meteor unusually fast or slow? Did it have a wake? If the meteor was bright, did it flare or break up, or leave a train of particles that lasted more than a quarter of a minute? Positive answers to these questions should be included in the comments column of a meteor report. It helps if you discuss these special characteristics with other people on your team; this is a good reason for having a team in the first place. Chances are that bright meteors will have been seen by more than one observer, especially if you have two people in each quadrant. So discuss the meteor: What things made it appear different from the others? What special notes should be recorded?

5.7.7 Plotting

Although tracing the path of each meteor you see onto a star chart is a helpful process when trying to monitor several minor showers at once, it is not necessary and is even detrimental to a successful counting program. The disadvantage to plotting is that it takes your carefully dark-adapted eye away from the sky for 15 to 30 seconds after you see a meteor and focuses your attention onto the illuminated sky map on your lap. Accurate plotting takes a lot of practice and continuing use of the skill to maintain quality. Because it takes time away from actually seeing meteors, those who undertake plotting do so knowing that their results cannot be compared directly with those who do not plot.

However, this word of woe is offset by the advantage of a plotted record of each meteor which is of value in determining radiants. If the night is especially busy, you might try having an associate sitting next to you with the map, and describe to that person the beginning and ending points of each meteor. 'Started near Alpha Herculis and ended close to Polaris' might work, though more detail is desirable (such as distance and position angle from the start and end stars). The plotter can then map it for you. N.B.: remember to mark the meteor's ID number at the start of its path on your plot, putting an arrow at the other end.

Normally, each observer is supplied with a single chart on which all his or her meteors for the night are plotted. If the shower is heavy, change forms after a specific time and note carefully when this was done. A gnomonic projection is normally used so that the plotted meteor is a straight line on the chart. (This type of map projection ensures that any line plotted on it is a great circle. The equator and meridians of longitude and right ascension are great circles; circles of latitude and declination are small circles. These charts are available from the International Meteor Organization (Appendix V).)

5.7.8 Observing meteor storms

Occasionally we are treated to a special conflagration of solar system debris after the Earth makes a close approach to some comet. The Leonids, the best known example of a meteor storm, stormed in 1799 and caused great panic in the nineteenth century, particularly in November 1833 and 1866, when their vast numbers terrified people. Those who were fortunate enough to witness the storm of November 1966, which produced hundreds of thousands of meteors on a single night, shared this spectacle with earlier generations.

On 9 October 1946, a storm resulted when Earth crossed the path of debris from Comet Giacobini–Zinner. The annual shower is called the Draconids, and these meteoroids have lower than usual density compared to those in other streams. But in 1946 this 'Giacobinid' display was so completely out of character that it surprised many observers. For 5 hours on that memorable night, and in spite of a bright Moon, some Canadian observers under Isabel K. Williamson counted over 2000 meteors. Since they observed only through specially made rings that allowed viewing of only selected areas, the count was far below what the actual was.

In 1985, a moderate Giacobinid shower with a maximum of 200 meteors/hour was reported by Y. Yabu of Japan at 0940 UT, a few hours before the Earth crossed the orbital plane of the main stream. Other observers from Japan also observed moderate counts for a brief period (IAU *Circulars* 4120 and 4124). Daylight radar results from the United Kingdom also indicated significant activity. These meteors must have been moved off the orbital plane by some perturbation.

Full meteor storms are extremely rare, but observers should always be ready, as many were for the moderate shower of 1985. In 1846, Comet P/Biela (1832 III) split into two comets, and was observed as a pair of comets in 1852. It was poorly placed during its 1859 apparition but the next one, in 1865–66, should have been favorable. However, no trace of it could be found. Then, at the time of its projected return in 1872, a storm of meteors appeared, described thus by Gerard Manley Hopkins, a famous English Victorian poet and observer:

> 'Great fall of stars, identified with Biela's Comet. They radiated from Perseus or Andromeda and in falling, at least I noticed it of those falling at all southwards, took a pitch to the left half-way through their flight. The kitchen boys came running with a great to-do to say something red hot had struck the meatsafe over the scullery door with a great noise and falling into the yard gone into several pieces. No authentic fragment was found but Br. Hostage saw marks of burning on the safe and the slightest of dints as if made by a soft body, so that if anything fell it was probably a body of gas, Fr. Perry thought. It did not appear easy to give any other explanation than a meteoric one'.
> (G. M. Hopkins, *The Journals and Papers of Gerard Manley Hopkins*, ed. H. House and G. Storey. London: Oxford University Press, 1959, 227–228.)

Observing the passage of the Earth through a dense concentration of meteor particles, like the Giacobinids of Comet Giacobini–Zinner in 1946 or the Leonids 20 years later, is a very rare treat. But even the regular annual showers can generate surprises. In 1982 the Lyrids produced a brief burst of 80 meteors per hour (N. MacLeod III, private communication), and in 1991 observers in Japan recorded a phenomenal Perseid display: one group, near Kiso Observatory, observed individual hourly rates during a three hour period of 64, 352, and 62 (IAU *Circular* 5330). Although this astonishing rate was missed by visual observers in Europe and North America, who were in daylight at the time, it was corroborated by amateur radio operators (IAU *Circular* 5345). The Perseids reprised this show in 1992 and in 1993.

During a major storm, if you try to observe the entire sky, or even a quadrant of it, you will not have a hope of making an actual count of meteors seen.

Some groups assign observers to small sections of the sky bounded by circles cut from cardboard. A well-defined asterism (an easily recognized grouping of stars that may not be a constellation) can also be chosen (and recorded) and used for making limited-area counts. This has the advantage of not applying blinders to the observer. Instead, you see much of the sky but concentrate for counting purposes on a limited area. The counting period can also be reduced.

These procedures offer a controlled way of making counts, but at the cost of missing some of the rare and grand spectacle. Besides, you will still not have the time to estimate the magnitude of each meteor; all you will be doing is counting. In reporting meteor storm observations, you should include an explanation of what procedure you followed on your report form.

During a meteor storm there is one other phenomenon to look for: faint glows may be visible in the direction of the true positions of the radiant and 'anti-radiant' (which may differ from their apparent positions by many degrees; IAU *Circular* 5840 corrects the prediction of Rao, 1993). These glows are thought to be sunlight scattered by tiny meteoroids still in their orbits in space, outside Earth's atmosphere.

5.7.9 Minor shower and 'sporadic' nights

Meteors can be observed any night of the year, and counting them on a night not dominated by a major shower can be especially rewarding. With experience you will have little difficulty assigning meteors to one of several active radiants, determining true sporadics, or possibly uncovering evidence of an unknown shower.

Before you begin, find out which showers may be within about a quarter strength of maximum and spend some time acquainting yourself with the positions of their radiants in the sky. On such relatively quiet nights, plotting is a useful approach, for it enables you to determine more accurately the shower membership of each meteor.

A note about 'Sporadics': we mark a meteor as sporadic when we cannot identify with reasonable certainty a known shower to which it might belong, or if it belongs to a shower in which, on a particular night, we have no real interest. The term is therefore used somewhat loosely, more as an observational note than a physical description. Perhaps a better term is 'nonshower' meaning not related to any shower being observed that night. There are probably no truly sporadic meteors; even one that might appear to be sporadic probably belonged to a stream that has long since dissipated.

If you suspect you may have uncovered a new radiant, do not announce your finding prematurely; there are so many suspected radiants and showers that occur once and never again that 'announcing' new shower finds is risky. If you strongly suspect that you have something, discuss it first with a few colleagues and others who are familiar with meteor studies, and plan to watch the sky the following year. Always stress that your results are provisional. If the subsequent watch does not confirm the new shower, that may mean that your new shower does not exist, or it may mean that the shower did not perform the second year and will return some other year, or it may mean that it was a shower that appeared but once. If you observe meteors often, you will likely uncover unexpected activity from time to time, which is precisely what makes meteor observing such a rewarding activity.

5.7.10 Fireballs, trains, and electrophonic sounds

Since they can attract such wide public attention, fireballs offer a useful way to introduce people to meteors. Although you may find it annoying that bright fireballs are so often misconstrued as UFOs, a little public education could turn faulty logic into a real interest in meteors.

A fireball is usefully defined as any meteor of magnitude −4, the brightness of Venus at its best, or brighter. Unless a fireball is at least magnitude −8, meteoriticists assessing the chances of its survival through the atmosphere would not take it too seriously. The report form that we suggest you use for reporting fireballs was designed by Dr. Peter Millman of Canada's National Research Council and is simple to use. This report form as well as one from the Royal Astronomical Society of New Zealand are reproduced in Appendix II.

Whether you use the form or not, the International Meteor Organization (Appendix V), which collects report of fireballs, requires the following information to assist searchers attempting to recover freshly fallen meteorites. The observational details should be reported accurately and as precisely as reasonable:

(1) The observer's name, address, and location of the observation.
(2) The direction the observer was facing for the observation.
(3) The path of the fireball: altitude and azimuth, position relative to known stars or geographic features, or any other method that provides a reliable description of the full path.
(4) Other descriptive details: did the fireball break up? Were sounds − rumbling, swishing, or crackling are possible − heard, either during or within a few minutes after the sighting? Was the meteorite itself seen?

Some time you might learn about fireballs others have seen. Your job is now to be a reporter with a responsibility to interview the observer and turn his or her inexperienced sighting into a report with scientific value. For instance, the observer may begin by telling you that the meteor was as bright as a street lamp, which of course is meaningless. But by careful questioning, you may be able to determine the brightness relative to street lights of known type and distance or to stars in the sky, the points of appearance and disappearance, and the direction of travel. If the fireball is comparable to the quarter

Moon in brightness, there is a chance it survived its plunge through the atmosphere, and can be recovered. Where a single report is of limited value, reports collected carefully from several observers at different locations over a wide area can be useful in locating possible meteorite falls. Fireball reports, singly or a collection, should be submitted to the Fireball Data Center of the International Meteor Organization.

Fireballs often generate long duration trains. Record data about the train including its duration (with unaided eye and with optical aid) and color. Sketches or photographs [Figs. 5.5] made over time will show the changes in shape as the train expands and drifts due to winds in the upper atmosphere. Binoculars or a telescope will allow detailed study of the fascinating changes that occur in these trains. With large binoculars, Stephen Edberg has seen clear indications that a train is an expanding cylindrical shell of material.

On rare occasions sounds may accompany the appearance of a fireball. A fireball that is the parent of a meteorite may be followed minutes later by the rumble of the meteorite's sonic boom during its supersonic passage through the atmosphere.

Even more rarely, swishing or crackling sounds may be heard *during* the appearance of a fireball or satellite re-entry. These sounds apparently travel at

Figure 5.5   This 45-second exposure made on Konica SRV 3200 film shows the expansion and motion of a fireball's train on 12 August 1993. Photograph taken at the Polaris Observatory Association site at Lockwood Valley, California by S. Edberg.

the speed of light, or at least their exciter does. Keay (1985) suggests a simple detector system for Very Low Frequency (VLF) radio waves that can be used to monitor observed fireballs. He reports (1991) that Japanese researchers have detected Extra Low Frequency (ELF) and VLF electromagnetic radiation from fireballs and that the cut-off magnitude for sustained electrophonic sounds is −9, while fainter fireballs with bright bursts may briefly generate sound. Thus, there is evidence that some low frequency radio waves generated by a fireball can be transduced to sound by natural objects.

5.7.11 Re-entering satellites

There are thousands of satellites, spent booster rockets, 'astronaut's gloves', and other baggage that now orbit our planet, enough to make the area just above our planet a junk yard. Chances are, you will see one of these bodies re-enter the atmosphere some day. They are not that difficult to identify, or at least to suspect, since their orbits around the Earth are vastly different from the solar orbits of meteoroids. Compared to the speed of a natural meteor, a re-entering satellite is slower, the 'unnatural' meteor lasts much longer, and often can be seen to fluctuate in brightness as it breaks up. We have seen re-entering satellites on several occasions; they have lasted for 30 seconds or more, refusing to disappear as they traveled at leisure across the sky, their paths sometimes exceeding 90 degrees.

Although this book does not have a satellite re-entry form, you can use the fireball report, noting your observation as a 'suspected satellite re-entry'.

5.7.12 Telescopic meteor observations

Counts of telescopic meteors made simultaneously with those obtained without optical aid can also be useful. Note your telescope aperture, field of view, and magnification. For each meteor, note the approximate right ascension and declination (or altitude and azimuth) of the center of the field of view. It has been our experience that even on nights of major shower activity, the numbers of telescopic meteors are not unduly high.

Variable star observers, who observe specific areas of sky at frequent intervals, are in an ideal position to add a telescopic meteor project to their observing program. Comet seekers can do this work as well, although you should be prepared to record reasonably accurate field centers and paths if possible for the meteors you see. Appendix II includes a telescopic meteor report form.

Telescopic meteor showers are showers whose average brightness is usually 4 or fainter. The Alpha Lyrids last from 9 to 20 July. Although the naked eye rate is usually no more than 2 per hour, through binoculars observers have reported as many as 30 per hour, though half that number is more typical. The June Boötids, originating from Periodic Comet Pons–Winnecke, last from

27 June to 5 July and have meteors averaging 5th magnitude, although at the shower's discovery on 28 June 1916, Denning recorded many bright meteors.

5.7.13 Recording data

Writing: Using copies of one of the forms supplied, simply write down, or have the recorder write down, the appropriate information for each meteor.

When you make a written record of a meteor, you must use a flashlight, usually red, so that you can see what you are writing. But even red light has some influence on the ability of the eye to see faint objects and deliberately varying the sensitivity of the eye through pauses to record meteor data is not good procedure.

Tape: Because you can superimpose your voice with audio signals from radio time signal stations and since it does not take your dark adapted eye away from the stars, tape recording is a method preferred by many observers. The only problem is that your post-observation work is increased, since you must now listen to the tape and copy all the information on to a report form which for a long observing night can become tedious. A microphone with an on-off switch allows you to tape only when a notation is made. Make sure that your recorder has enough electrical power – batteries fail in cold weather – and if the temperature is below freezing the recorder may not work properly in any case.

Counter: If you are observing alone and all you want is simple hourly rates, you can use a gate counter, a finger-keyed supermarket hand adding machine, or a golfer's stroke counter that you press every time you see a meteor. You will get hourly rates this simple way, but not the other information which is also useful. The advantage of the counter, when pressed for each shower meteor you see, is that it allows you to observe for a specific goal of determining shower rates, and not worrying about anything else.

5.7.14 Report forms

We offer a choice of two totally different forms. The first, designed by Dr. Peter Millman and used with his permission, is arranged so that you can record each meteor you see, while the second is set up simply to record hourly totals. These two forms represent different philosophies of recording meteors, the first paying more attention to the properties of each meteor. You will find the two meteor forms, as well as the fireball form, in Appendix II. Use the form with which you are more comfortable. Results by single observers require no further reduction and should be submitted to meteor observation societies

(Appendix V) 'as is'. Such data can be incorporated directly into a single observer database.

5.7.15 Making the observations

5.7.15.1 Single observer

Watching meteors alone is a personal communion with nature in our neighborhood. Where a group observing project introduces its members to a shower, an observer working alone enjoys a gentle evening out on Earth's front porch, watching visitors pass by as meteors fall. Choose an observing site that promises to permit a relatively high number of meteors to be seen.

Solo watching has the advantage of not needing too much planning. Depending on what shower you are watching you might see 50 meteors per hour or more under an ideal sky if you are vigilant and lucky. Individual observations can be combined without effort into certain data bases such as those maintained by the Meteors Section of the Association of Lunar and Planetary Observers or the American Meteor Society.

First, remember that you cannot hope to cover the whole sky. Choose a part of the sky in a direction away from city lights, one that promises the greatest number of meteors. This is not necessarily the direction of the radiant of whatever shower you are trying to observe, since meteors can occur almost anywhere. A lone observer should try to watch some 60 to 90 degrees from the radiant.

The exception to this is the period from 20 July to 14 August, when meteors from two major radiants are falling, the Delta Aquarids and the Perseids, and the minor radiants of the Alpha Capricornids, the Kappa Cygnids, and others, at the same time. During this period you should face the south if you want the best chance of distinguishing a meteor belonging to one shower from a meteor hailing from another shower. Meteors can be recorded either onto one of the standard meteor report forms or onto magnetic tape to be transcribed later.

Do not be overly ambitious with recording your data; you'll miss meteors. Consequently, it is not a good idea to plot each meteor if you are watching alone for the purpose of hourly counts. A tape recorder allows count data to be stored without removing your eyes from the sky.

Watch your area carefully for at least one hour, and two or three hours are better. Observation periods shorter than that will generally not be valid statistically.

For each meteor, you need to record the visual magnitude, the meteor's shower membership, and any comments such as 'unusually fast' (abbreviated 'ufst') or 'left three-second train' (abbreviated '3st'). (A personal set of standard abbreviations will save writing time when your eyes are off the sky, and of course, there would be no problem at all if you used a tape recorder.) You are after statistical data, so that you do not have to write down the time to the second of every meteor you see as well as the area of sky you are observing.

A special caution is needed about magnitudes. Many observers, especially new ones, tend to overestimate the brightness of a meteor. You are watching an object appear suddenly, move across some stretch of sky, then disappear, and then you are supposed to compare it to the brightness of a quiet and steady star. Be careful! Also, try to choose comparison stars that are close in altitude to that of the meteor, so that atmospheric extinction will not affect the estimate you give.

5.7.15.2 Group observing

Some debate has surrounded the best way of observing in groups, with conflicting suggestions that observers should each record their own meteors or have a central recorder. If groups are organized the first way, then each observer will have a separate record sheet to write down the magnitude, shower membership, and any comments about each meteor seen. If the gathering employs a central recorder, all meteors are written down on one sheet, and only one master list is kept. This system respects the fragility of the human eye and its desire not to have unwanted light thrust upon it. Radio time signals or a central clock that can be used by all observers should be provided.

Overlapping observing areas slightly is a good idea, so that faint meteors appearing at sector edges will have less chance of being ignored. Since we are interested in meteor counts per observer, meteors seen by more than one watcher should be recorded by each of them. If a group count is kept, and meteor 41 is seen by two observers, it should thus appear as two reports, either from the individuals or on the master list (which will indicate the observers).

Team of 5

In this setup, one observer takes each cardinal direction, with the fifth as recorder. Observe together in hour-long periods, with reasonable time allotted for breaks; don't try to have the observers watch for too long without a rest!

The central recorder is notified each time someone sees a meteor by the traditional call of 'TIME!' The recorder then assigns a number, which the observer records if he or she plots the sighting, and then asks for the magnitude, shower membership, and any comments.

Team of 10

With this many people, you have several possibilities. You can have all 8 observers and two recorders working at once, with the sky divided into the above-45 and below 45-degree parts of each quadrant. This may be the best way if you plan to observe for only one or two hours. For longer sessions, you may plan to rotate assignments, so that half the team is observing and the other half resting at any time.

136

The third possibility is somewhat unconventional but can be interesting and fun. Divide the group into two teams at locations differing by some 65 to 80 km (40 to 50 miles), with instructions to record and plot carefully the time and path of any meteor first magnitude or brighter, and later try to determine through trigonometric calculation (see Meteor triangulation in Section 7.1.4) the height and distance of these meteors.

Team of 15

A good way of observing with this many people is to organize a set of three groups of four for observing and one group of three for recording. A schedule should be prepared that will allow each observer one hour on, followed by one half hour off. The hour of observing is divided into two parts, the first for observing the upper part of a quadrant, the second for the lower, thus keeping the mind alert through changing position. A typical schedule is given in Table 5.2, each letter standing for an observer.

The advantages of such a system are easy to see. First, the observers get equal opportunities to rest; they are on for an hour and off for a half hour. Second, the switching of observers from upper to lower halves of a quadrant adds some element of randomness to the counts, although the most astute observers will still always be in the same quadrants.

For some amusement, the Royal Astronomical Society of Canada's Montreal

Table 5.2. *Sample observing schedule*

Local time	Quadrant half	North	South	East	West
2000	Upper	A	D	G	J
	Lower	B	E	H	K
2030	Upper	B	E	H	K
	Lower	C	F	I	L
2100	Upper	C	F	I	L
	Lower	A	D	G	J
2130	Upper	A	D	G	J
	Lower	B	E	H	K
2200	Upper	B	E	H	K
	Lower	C	F	I	L
2230	Upper	C	F	I	L
	Lower	A	D	G	J
2300	Upper	A	D	G	J
	Lower	B	E	H	K
. . . and so on.					

Centre had a special citation for observers who spotted meteor 100, 200, and so on, during a meteor watch. These privileged people got to join the 'Order of the Hole of the Doughnut'.

Special considerations
In any amateur meteor watch, where the observers do not have the same eyesight, the same experience, the same motivation, or the same understanding of what is happening, some problems obviously appear. The most common:

Facing the radiant: Place the most experienced observer toward the radiant, or certainly within 90 degrees of it, so that shower versus nonshower meteors can be determined more precisely. During the major summer showers, all the experienced observers should face east and south, again to determine precisely to which shower a meteor belongs.

Different levels of experience: How does a team operate scientifically when its members do not all have the same degree of expertise? We see no problem with this, so long as the degree of reliability is shown clearly. On a separate sheet at the front of a meteor shower report, indicate, on a scale of 1 to 5, (5 best), your opinion of the experience of each observer. A person on his or her first watch might be assigned a 1 or 2, while someone with more maturity, and a few more meteor showers, would get a 3 or 4. The experienced observer would get a 5. This way, the data can be reduced with a realistic idea of which data are most reliable. Don't assume that the observer with the highest quantity of meteors is necessarily the one with the highest quality.

   Watch carefully the patterns of each observer. See if, in your judgment, meteors might have been missed, or if magnitude estimates are consistently too high or too low. See if the observer has a clear understanding of radiants, and is tracing paths properly. As recorder, you can keep an eye on such things, for surely you will see for yourself some of the meteors the others are reporting.

The High Counter: No matter which quadrant he or she is assigned to, this observer will always have the highest numbers of meteors. There are three possible reasons: (1) This observer likely has more alert eyes and mind than the others, and is catching more objects; (2) An observer sometimes catches fainter meteors than the others; or (3) A vivid imagination is behind high counts. This last condition might be difficult to detect. If the high counter is a beginner, the eyes may wander off into unassigned areas. Also, if the beginner's meteors are faint, some of them may be phantoms. Usually, the highest counters are the most experienced, and it is wise to assign the two highest counters to opposite quadrants.

Reducing group data

The process of preparing your data for close scrutiny is laborious, but always instructive. You should list the results of your observing sessions by each type of information that was recorded. First, tabulate the meteors seen by each observer per hour, a process that will help you determine the capabilities and reliability of each person on your team. It is simplest to make starting and ending times on the Universal Time hour. If the observers record their data individually in the first place, then you are saved this aspect of reduction. After the meteors are listed in this way, they should be carefully totalled.

Reduction by observing position: Rather than offering any special value to science, this listing acts as a control, helping to check the visual acuity and accuracy of your observers.

Reduction by magnitude: Record the night's work graphically so that magnitudes are listed across the top of your page, and time listed down the page. This type of listing can show some unexpected turns in what you think a meteor shower will offer. Did this year's shower show more fainter meteors than last year's? When the total number increased after midnight, was the increase reflected across the board or only on the fainter meteors? Compare this with the observer analysis to see if your answers to these questions are affected by differences in observer perception.

Reduction by shower: When you plot shower against time, you have a chance to get a general picture of the relative strength of a shower. This reduction is especially useful when several showers are active at the same time, as happens in July and August. It also provides an opportunity to evaluate the understanding your observers have of meteor radiants.

Sample group reduction: This is a sample of the reduction of a major meteor observing session that was held on 12/13 August, 1966. The work was done carefully by Isabel K. Williamson, a prominent meteor observer in Canada who has introduced hundreds of people to the special joy of meteor gazing. In the days before home computers, this work was time-consuming, and this particular reduction took several nights of dedicated work to complete.

> The perseid meteor shower
>
> (reprinted from *Skyward*, October 1966, the monthly newsletter of the Royal Astronomical Society of Canada, Montreal Centre)
>
> On the invitation of Mr. Granger Robertson the Centre's meteor team observed the Perseids from his summer home at Ste. Margeurite, Qué., on the night of the maximum, August 12/13.
>
> There was the usual overcast sky when we left Montreal. (As one of the team remarked, we wouldn't feel comfortable if the sky were clear when we

set out on one of these jaunts.) We drove through the usual rain shower. We arrived at our destination and determinedly went about setting up the equipment, trying to ignore the heavy clouds. We went indoors for the usual briefing. At 9:45 p.m. EDT, one or two stars were visible and we decided to 'go through the motions' for the benefit of newer members of the team. Light rain was actually falling at 10 p.m. when we took up our observing positions but a few stars were still visible. The first meteor was called within the first five minutes and two more in the next five, which encouraged us to continue. Then the sky began to clear. By 11:30 p.m. there wasn't a cloud in the sky and we enjoyed perfect observing conditions right through until dawn. In six hours of observation we recorded 906 meteors, thus breaking our record for all showers except the famous Giacobini-Zinner shower of 1946. It was a fantastic night.

During the night we saw passages of Echo I, Echo II and Pageos I but were much too busy with meteors to record the times of passage. We did sing 'Happy Birthday' to Echo I which we had observed from Montgomery Centre on the night it was launched exactly six years before.

We have now completed an analysis of our meteor team's observations on the night of August 12/13 at Ste. Marguerite, Qué., and have some interesting statistics to report. Table I shows the distribution of meteors recorded at half-hour intervals for the four observing positions.

Table I. *Distribution by Time Interval and Observing Position*

E.S.T.							
From:	To:	North	South	East	West	½ Hour	Hour
21:00	21:30	4	—	3	—	7	
21:30	22:00	10	3	3	1	17	24
22:00	22:30	18	9	9	3	39	
22:30	23:00	29	10	8	13	60	99
23:00	23:30	26	28	13	14	81	
23:30	00:00	42	23	8	11	84	165
00:00	00:30	57	22	20	19	118	
00:30	01:00	17	34	13	16	80	198
01:00	01:30	53	29	15	17	114	
01:30	02:00	38	27	11	21	97	211
02:00	02:30	25	31	14	20	90	
02:30	03:00	53	39	13	14	119	209
Totals		372	255	130	149	906	906

The low count during the first hour of observation is attributable to the poor observing conditions. For the rest of the night skies were exceptionally good and the table shows the expected increase after midnight. Three half-hour periods were very busy. The meteors came in bunches and there were some frantic moments – for instance, the five-minute period from 00:10 to 00:15 when 30 meteors were recorded.

Table II. *Distribution by Time Interval and Individual Observer*

Position	North			South			East			West			
Observer	W	B	RA	D	S	JO	JU	L	C	G	N	RO	Total
From: To:													
21:00 21:30	0	4	—	0	0	—	0	3	—	0	0	—	7
21:30 22:00	3	—	7	2	—	1	0	—	3	1	—	0	17
22:00 22:30	—	13	5	—	4	5	—	4	5	—	2	1	39
22:30 23:00	5	23	1	3	7	—	1	7	—	6	7	—	60
23:00 23:30	14	—	12	12	1	15	9	—	4	8	—	6	81
23:30 00:00	—	24	18	—	10	13	—	5	3	1	6	4	84
00:00 00:30	13	43	1	10	11	1	6	14	—	7	12	—	118
00:30 01:00	7	1	9	14	2	18	5	1	7	11	1	4	80
01:00 01:30	—	37	16	—	13	16	—	9	6	1	9	7	114
01:30 02:00	8	30	—	9	18	—	5	6	—	4	17	—	97
02:00 02:30	11	—	14	14	—	17	8	—	6	6	—	14	90
02:30 03:00	—	32	21	—	22	17	—	6	7	—	6	8	119
Totals	61	207	104	64	88	103	34	55	41	45	60	44	906

Note: Observers in Preceding Table

W Bill Warren
B Si Brown
RA Boyd Ramsay
D Louis Duchow
S Arno Schmidt
JO Carl Jorgensen

JU Walter Jutting
L David Levy
C William Cullinan
G Charles Good
N Leo Nikkinen
RO Granger Robertson

Table III. *Magnitude Distribution by Individual Observer*

Magnitude	Brighter than 1st	1.0 to 1.5	2.0 to 2.5	3.0 to 3.5	4.0 to 4.5	5.0 and fainter	Total
Observer							
Brown	13	9	22	22	48	93	207
Ramsay	5	19	23	26	18	13	104
Jorgensen	11	19	30	27	14	2	103
Schmidt	5	13	21	18	21	10	88
Duchow	13	13	15	19	3	1	64
Warren	5	8	26	9	9	4	61
Nikkinen	2	13	13	16	13	3	60
Levy	4	7	7	14	17	6	55
Good	3	8	13	17	4	—	45
Robertson	1	9	12	17	3	2	44
Cullinan	—	10	14	10	4	3	41
Jutting	1	9	3	8	10	3	34
Totals	63	137	199	203	164	140	906

For ready comparison Table III is arranged according to the number of meteors recorded by each observer. It will be noted that no one had the corner on meteors 'brighter than 1st'. Duchow caught as many as Brown. For meteors ranging from 1st to 4th magnitude, Brown's count was neither the lowest nor the highest. For the fainter meteors, though, his count far exceeded that of any other observer. Si Brown, then, is one of those fortunate individuals endowed with exceptional eyesight. If his count for the fainter meteors were reduced to the average recorded by the observers, his total count would be cut in half.

5.8 Meteor photography

Meteor photography is one of the simplest and one of the most frustrating areas of astronomical photography. While it is an area in which even a beginner can successfully participate, the frustration comes from the need for luck to have a bright enough meteor flash through the camera field of view.

At its simplest, meteor photography requires only a camera with a fast (low f/number) lens whose shutter can be opened for indefinite periods (using a B or T setting) and a sturdy support to aim it at the night sky. Even the more elaborate techniques described below require little more than this basic apparatus.

5.8.1 Lens and film selection

The choice of camera lens and aperture is not as obvious as might be initially assumed. Sidgwick (1982) discusses selection in some detail. For comparing the relative efficiencies of camera lenses for meteor photography the expression $F/(f^2)$, where F = lens focal length and f = focal ratio (lens focal length/lens aperture), given by Sidgwick, should be evaluated. For older lenses with severe image degradation at the edges, the size of the useful image plays a role too.

We can understand the meaning of this expression by rewriting it to its equivalent: $(D^2)/F$, where D is the lens diameter. The meaning is now clear: a greater aperture collects more light, while a longer focal length decreases the efficiency by making a larger image that spreads the light over a larger area, rendering it fainter on the emulsion. Since increasing aperture gathers light more rapidly than focal length spreads it, the message is that a large aperture and small focal length are the most efficient. A 50-mm f/2 lens ($D = 25$ mm) will catch more meteors than a 28-mm f/2.8 lens ($D = 10$ mm) because it is 3.5 times more efficient in light gathering power. This gain in efficiency holds even when the sky area covered by the lenses is considered, because the wide angle lens covers only 2.8 times the area covered by the 'normal' lens.

For 35-mm cameras that are now available virtually everywhere the choice is simple: choose the fastest (lowest f/ratio) lens available. For the same negative size, the field of view of a lens is of secondary importance because the aperture determines the amount of light collected and therefore the faintness limit of the meteors recorded. Wide angle lenses tend to have small apertures and thus are useful only for capturing fireballs on film, and not very well most of the time [Fig. 5.6]. Standard lenses, $F = 50$–55 mm, $f/1.2$–2 are in general the most efficient for capturing meteors. Comatic images off-center can be a problem with fast lenses but can be reduced by closing the lens one or two stops. Of course, doing this will also reduce the meteor capture rate.

The choice of film is fairly simple: the more sensitive the emulsion the more likely a meteor will be captured. The speed rating should be ISO 400 or higher. The only concerns are the grain size in images and the effect of general sky brightness (including twilight, aurora, zodiacal light, the Milky Way, and light pollution).

The grain size generally increases with emulsion speed and can be disturbing when enlargements are made. High sensitivity films, especially when combined with fast lenses, can have images washed out by overexposure to general sky illumination.

Sky illumination in modest amounts can act to sensitize the film for fainter meteors. Other hypersensitizing techniques are also useful, and under the

Figure 5.6 A 1988 shadow-casting Perseid fireball captured by a reflective all-sky camera (effective focal length about 8 mm, f/4). Note the difficulty in distinguishing the fireball on the sky: the fireball is near the base of the center leg. Photograph by S. Edberg.

photographer's direct control, for meteor photography. However, because these techniques reduce low intensity reciprocity failure for both long and short exposures, the emulsion's sensitivity to the background sky will also be increased and sky fog, reducing the apparent contrast between a captured meteor and the background sky, will be a problem with long exposures. Of course, this is a problem for any film making a long exposure for meteors. Reynolds and Parker (1988) have demonstrated sensitivity increases of 2 to 2.5 for planetary exposures, which roughly match typical meteor durations, on hypered Technical Pan.

Stephen Edberg's experience indicates that a meteor photographer can expect to capture 2–3 meteors per camera per night during such major showers as the Geminids and Perseids using 50-mm $f/1.8$ lenses and ISO 400 emulsions under good dark-sky conditions.

5.8.2 Aiming

For meteor photographs the camera should be aimed about 45 degrees above the horizon and 45 degrees from any active meteor radiants (or from the most active one). This allows the camera to monitor a large amount of the atmosphere where meteors occur. Cameras aimed lower may not be as effective capturing meteors because of the atmosphere's absorption of light near the horizon.

5.8.3 Using stationary support

Once mounted on a solid support and aimed up, the camera shutter is opened and locked with a cable release until a meteor flashes through the target area or fog due to sky light (natural or artificial) threatens to ruin the photo by washing out everything. For the fast emulsions discussed earlier, exposures of a few minutes to as long as an hour can be made depending on the local sky brightness conditions, and must be determined experimentally for each site used and sky condition encountered.

5.8.4 Using a clock drive system

Instead of a stationary support, cameras can be mounted on their own equatorial mount or piggyback on a telescope. This allows constellation or Milky Way photographs to be made even if no meteors are captured.

Several cameras can be mounted together on the same drive system if desired. This allows more area of the sky to be covered. (Beginning photographers may be surprised by how little of the sky fits in the field of view since a 50-mm lens typically covers only 47 degrees across the diagonal of a standard 35-mm frame.) The cameras should be aimed so that their fields of view overlap slightly at the edges. 'Ball heads', available from camera stores, are very useful in aiming and orienting cameras. As with other astrophotography, a sturdy, vibration-free drive system is desirable.

When using several cameras, it is convenient to mount all the cable releases together [Fig. 5.7]. A hinged device can be built to trip all the shutters simultaneously, and even lock them, but a compressible rubber pad may be necessary to absorb the different amount of travel different cameras may require in their shutter release mechanisms.

It is worth investigating the different types of cable releases available. The cheapest, cloth covered types may fail at critical times. One-hand and two-hand lock/release mechanisms are available; Stephen Edberg greatly prefers the one-hand lock/release cable releases. Two-hand mechanisms require the photographer to hold the plunger in with one hand while tightening a thumbscrew with the other, an inconvenient and potentially difficult operation, especially with gloves on.

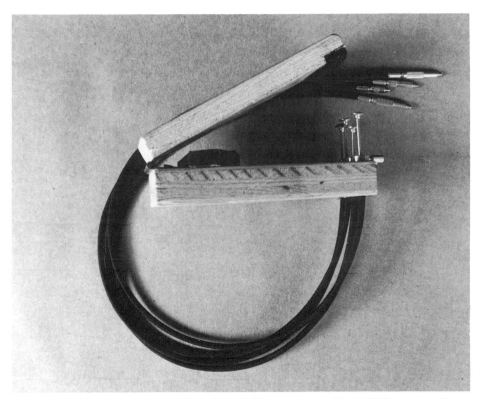

Figure 5.7   This home-built cable release holder constructed by J. Edberg permits four camera shutters to be opened simultaneously. Commercial units with one plunger for two cameras can be found on the market.

145

5.8.5 Electronic image intensifiers – meteors and television

An image intensifier with a light amplification factor of several thousand allows meteors much fainter than the photographic limit near magnitude 1 to be recorded. Because background skylight, both natural and artificial, is also amplified, photographic exposures through an intensifier can be badly over-exposed in seconds! Intensifiers are useful though, for video tape recording of meteors when combined with a recording of radio time signals. One can then record and later count meteors in a well-defined sky area over known time intervals. The intensifier and video camera combination can be mounted either stationary or on an equatorial mount. The addition of an objective prism or diffraction grating allows meteor spectra to be recorded.

Peter Manly of the Saguaro Astronomy Club of Phoenix, Arizona, has pioneered amateur techniques for observing meteors using television equipment and image intensifiers. Besides the strictly scientific benefits of this system, it has an added bonus for publicity: Manly's results have been broadcast on the evening news in Phoenix, so that a potential audience of hundreds of thousands could get an introduction to what meteors are and how much fun one might have observing them. Manly writes:

> Recent advances in the electro-optical industry have brought television and image intensifier technology within the reach of the amateur observer. There are now rugged, reliable, portable and relatively inexpensive television cameras, image intensifiers, and video tape recorders which are capable of recording large sections of the night sky down to fifth or sixth magnitude. The approach is to use a commercially available TV camera similar to the type sold to make home video tapes. Since such cameras are not sensitive enough to see meteors fainter than magnitude -1, an image intensifier and a lens with a wide field of view replaces the normal TV camera lens. A home video cassette recorder preserves the data.
>
> The system we have used yields a 40 degree (measured diagonally) field of view with a Tokina 12.5 mm focal length $f/1.3$ TV camera lens as the objective. If we used shorter lenses, the smaller apertures would limit the sensitivity of the system, and with a larger lens the field of view would be smaller. The image intensifier is a 17 mm diameter second generation microchannel plate device, similar to those developed for the military for seeing at night. It is basically a device which takes a faint image generated by the objective lens and increases its intensity. It is coupled to the TV camera with a relay lens, and its output image glows in green light.
>
> The performance of this system, which also includes a WWV radio played on the audio track, allows us to see stars to visual magnitude m_v 5 or 6 on a clear, moonless night. We can see meteors to about 4th or 5th magnitude depending on their velocity.
>
> Observing with this system is fairly simple. The camera is set up on a tripod and pointed toward the appropriate area of sky. It is best to use a dark sky site since the camera is sensitive enough to pick up the background from city lights. The tape recorder is run for the duration of the observation. The tough part comes during data reduction, when you must stare at the screen for as

many hours as you have recorded in order to find the meteors. *We noticed many meteors on review of the tape which were not seen during the observations.* This may be partly due to the fact that the observer was tired during the observations. One advantage is that when the observer becomes fatigued while reducing the data he can stop and rest. During live observations this cannot be done since it would mean missing some data.

The results so far have been encouraging. We tried the 1984 Orionid meteor shower and we were able to use two cameras in order to obtain triangulation of the meteor positions. Edited sections of the video tapes were also replayed by a local TV station news department.

Since Manly wrote these paragraphs, home video technology has progressed considerably, somewhat to the detriment of astronomical users. In particular, cameras with separate video cassette recorders have virtually disappeared. They have been replaced with one-piece 'camcorders'. This is advantageous in that only one unit must now be maintained at its operating temperature on a cold night, however, it is now more difficult to find camcorders with interchangeable lenses, so that a lens plus intensifier set can be used.

5.8.6 Meteor choppers

The addition of a rotating blade in front of a camera's lens allows additional data on any photographed meteor to be gathered. Knowing the chop rate and counting the number of breaks in the trail allows you to determine the duration of the meteor [Fig. 5.8]. Combined with the triangulation method discussed in Section 7.1.4, actual in-air velocities and decelerations can be determined. Choppers also allow longer overall exposures because they reduce incoming skylight by the ratio of their opaque area to the clear area.

To make a chopper, mount a propeller- or windmill-shaped piece of plastic or wood on a motor. The 'blades' should be wide enough to completely block incoming light from reaching the lens, otherwise a meteor trail would be bright–faint–bright–faint–bright . . . and so on, since light from the meteor would still be reaching a part of the lens and going on to the film. When the blades from a chopper completely cover the lens intermittently a meteor will be recorded with the desirable on–off–on–off–on . . . pattern. Note, however, that meteors leaving persistent trains will leave trails that are bright–faint–bright . . .

The motor should be synchronous or have a mechanical governor: for proper analysis the chop rate must be known accurately. To avoid having motor vibrations smear images, it is desirable to mount the chopper assembly independently of the cameras. Stephen Edberg's experience is that most clock motors do not transmit vibrations to cameras mounted on the same support. However, rapid rotators (60–90 revolutions/minute) have little torque for turning a heavy chopper. A synchronous record player motor should work well.

Figure 5.8   This mag. −4 Geminid meteor, to the right of the Big Dipper, was captured with a 24 mm f/2.8 lens and chopper rotating at 90 revolutions per minute with 10 blades on the rotor. Thus there are 900 breaks/minute, or 15 breaks/second. The 12 segments visible in the trail (on the original slide) indicate the meteor's duration was 0.8 second (12/15 second). Photograph by S. Edberg.

A chop rate of 12 to 60 breaks/second is desirable. This can be determined by multiplying the number of blades on the chopper by the motor's revolutions per minute (rpm) and dividing by 60: (slots) × (rpm)/60= breaks/second.

For more discussion of choppers see McKinley (1961), Sidgwick (1982), and Ridley (in Moore, 1963).

5.8.7 Meteor train photography

The photography of a persistent train is both challenging and useful. A camera with a fast lens and sensitive film must be available for immediate pointing at the train. Because of the low intensities, the photography of a meteor train must balance the need for sufficient exposure against the smearing caused by the expansion of the train and its bodily motion due to upper atmospheric winds. As a photographer, you must judge what the sufficient minimum exposure is for the train and then obtain a series of photos (changing the exposure as the train fades). Record the time and duration of each exposure. Studies of the train's evolution and motion are most valuable.

5.9 Radio methods of meteor observation

Studies of meteors by radio have been made by a variety of means since the 1920s (McKinley, 1961). With access to funding and equipment, professional astronomers actively probe meteors (in particular, their ionization trails) with radar. Amateurs are limited to more passive studies, but they still require an artificial source of radio transmissions.

This requirement is not as stiff as it might seem because standard FM radio or television transmissions can be used as the source (Lynch, 1992). At its simplest, FM meteor observing requires only that an FM radio or TV be tuned to a frequency free of local interference, i.e. all the transmitters on that frequency must not be directly detectable by the receiver. Choose a frequency at the low end of the FM band or in the range of television channels 2–6, the lower the better.

Short term (seconds long) signal enhancements will then be due to the reflection of signals off the ionized trail of a meteor (or possibly off the skin of an aircraft). (Enhancements due to solar activity and other effects on the ionosphere are possible but last longer and thus are distinguishable from the brief effect of a meteor.) Signals may be reflected by meteors from broadcasts as far as 1600 km (1000 miles) away; you may not see the meteors doing the reflecting because they may be close to the horizon and faint due to atmospheric absorption.

On the higher frequencies, hourly counts can be made 24 hours a day, if desired, indoors, and in all kinds of weather. Observations can even be tape recorded for later analysis, and more elaborate observing and recording schemes are possible (Suzuki *et al.*, 1976, Pilon, 1984). Similar work can be done at short wave frequencies (Setteducati, 1960).

An American Meteor Society bulletin by D. Meisel outlines the possibility of using aeronautical radio beacons as a source for reflection monitoring. Black (1983) has developed a system utilizing these beacons. Experiments using low power beacons and other 'passive' transmissions are continuing. Also, US Federal Communications Commission (FCC) regulations now allow the establishment of amateur radio beacons so that frequencies outside the FM-TV bands may be used for meteor monitoring. Thus, observing groups or individuals can set up their own systems if desired.

A great deal of useful information in radio amateur ('ham') magazines on the subject of radio scatter communications is available. Meteor observers may wish to team up with local hams to set up a meteor watch system. Short introductions to meteor scatter can be found in *The 1992 ARRL Handbook for Radio Amateurs* (p. 22.15), *The ARRL Operating Manual* (pp. 12.8–12.10) and in Bain (1957 and 1974), Greene (1986), and Owen (1986).

The Quadrantids, Arietids, Perseids, Leonids, and Geminids are popular showers among hams. Pocock (1992) reports on the 1991 Perseid activity and Harris (1967) reports on the 1966 Leonids.

Besides monitoring shower and sporadic meteor rates any time and in any weather, another useful project requires concurrent visual and radio observations. Although calibrating meteor visual magnitudes and radio detections may be complex, the calibration of meteor rates obtained by visual, photographic, and radio methods would be useful. Studies of radio signal strength and duration compared to visual magnitude and photographic duration can be made. The visual position in the sky may affect the radio detection. Numerous useful studies can be made, limited only by the observer's imagination. With renewed professional interest in meteor science, amateur studies using radio should take on greater importance.

6

The zodiacal light

6.1 Introduction and historical notes

In any book about the minor bodies of the solar system, the zodiacal light must play a unique part. While the zodiacal light, in fact, covers the whole sky, and has notable enhancements in certain directions, it is the triangular patch of light that so delicately climbs from the horizon into the sky along the ecliptic that most observers recognize [Fig. 6.1].

Although the zodiacal light has been a part of the sky for much of the life of the solar system, and certainly for as long as observers have been around, it has not been written about nearly as much as other aspects of the sky. It was first mentioned as a special phenomenon by Cassini in 1683. T. Brorsen called attention to the zodiacal light enhancement opposite the Sun in the sky in 1854.

A comprehensive discussion, including many illustrations of the appearance of the zodiacal light, appears in a report published in 1856 after the 'Expedition of an American Squadron to the China Seas and Japan, Performed in the Years 1852, 1853, and 1854, Under the Command of Commodore M. C. Perry, United States Navy'. Entitled *Observations on the Zodiacal Light, From April 2, 1853, to April 22, 1855, Made Chiefly on Board the United States Steam-Frigate Mississippi, During Her Late Cruise in Eastern Seas, and Her Voyage Homeward: With Conclusions From the Data Thus Contained*, the Rev. George Jones, A M, Chaplain, USN, provides verbal descriptions with hundreds of sketches of brightness contours of the zodiacal light pyramids in the sky, as well as a long discussion and his conclusions.

In 1884 Gerard Manley Hopkins wrote *Spelt from Sybil's Leaves*, a complex sonnet that begins with an intricate view of the sky after sunset. In its third line he touches on what David Levy believes is the zodiacal light, as a 'wild hollow hoarlight hung to the height'. Hopkins was an astute observer of nature. In his journals and poems he recorded his observations of natural effects both on Earth and in the sky, his goal being to uncover what he defined as 'inscape', in which a special pattern or order could be found in the simplest of nature's details. He studied things like the pattern of a leaf being copied on both sides, the shape of wind-driven snow, and the peculiar uniformity of the glows in the night sky.

Figure 6.1   The zodiacal light is a faint glow in the morning sky in August 1978. This is a 10 minute exposure on GAF 500 film made with a 35 mm f/2.8 lens. Photograph by S. Edberg.

A few months before Hopkins began work on his poem, the British observer G. M. Whipple (1884) published a report in *Nature* of a 'brilliant appearance' of the zodiacal light pyramid, 'the cone of light being exceedingly well defined'.

Since the glow is made by millions of particles spread through a huge volume of space, it is difficult to attribute changes in it to any intrinsic property of the light. It is more likely these perceived changes are the result of varying degrees of clarity of the Earth's atmosphere. Stephen Edberg notes that the explosion of the Indonesian volcano Krakatau (or Krakatoa) occurred in 1883 and blasted large amounts of dust into the stratosphere, causing spectacular sunsets for several years following the eruption. Could the volcanic dust and gases high in the air have caused the effect observed by Whipple? Observations of the zodiacal light after major volcanic events like those of El Chichón in the early 1980s and Mount Pinatubo in the early 1990s might provide the answer.

6.2 Astronomy of the zodiacal light

The zodiacal light is actually a complex of skyglow features, so named because it is mostly seen in the constellations of the zodiac, through which the planets pass. It is concentrated near the ecliptic, the intersection of the plane of Earth's orbit with the sky. It actually does not precisely coincide with that plane, but instead is tilted about 1.7 degrees from it. The light reflects the presence of a virtually limitless number of small, fluffy particles, each in the size range of 1 to 100 micrometers, and each orbiting the Sun independently. As sunlight scatters off this inhomogeneous mass of particles, they produce an elusive glow that can be seen by a dark-adapted observer in dark skies away from cities and not bothered by moonlight.

In fact, the glow from these particles found all over the sky will ultimately limit the performance of the Hubble Space Telescope as it studies extremely faint objects. Consider this analogy: trying to watch a movie in a lighted room is difficult so you turn down the lights. But if the room can't be darkened enough you lose the ability to distinguish the action on the screen. In the case of the Space Telescope, the zodiacal light is ultimately brighter than the faintest objects it could otherwise detect, thus washing out the faintest 'action' on the sky. The only escape from such a problem is to move the orbiting satellite beyond the asteroid belt (beyond about 3.3 AU), where the dust density falls dramatically.

The closer to the Sun, the more concentrated these particles become. Some zodiacal light particles come from the production of dust by comets as they pass through the inner solar system. In 1983 the Infrared Astronomical Satellite, abbreviated IRAS, surveyed the sky in infrared wavelengths [Fig. 6.2]. Its data, coupled with other studies of cometary dust production rates, determined that comets do not produce sufficient dust to maintain the necessary quantities of dust and that the dust rings associated with the asteroid belt between Mars

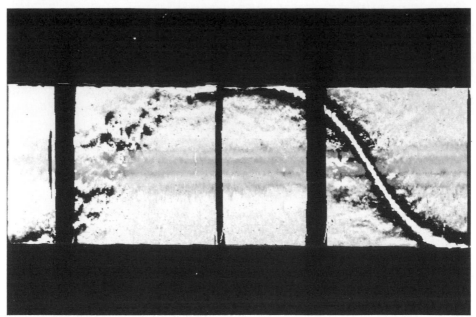

Figure 6.2   Dust in the solar system as recorded by the Infrared Astronomical Satellite (IRAS). In this map projection the three horizontal bands are zodiacal dust. The S-shaped curve is infrared emission by our Milky Way galaxy. Courtesy of NASA/JPL.

and Jupiter must contribute more than, and perhaps much more than, 50 percent to the particle cloud. It is interesting to note that some of the bands have been associated with the Themis, Eos, and Flora families of asteroids.

The particles in the cloud do not orbit indefinitely. The cloud is continually depleted by the gravitational sweeping of the cloud by the planets, the Poynting–Robertson effect (wherein the absorption and emission of light causes the smallest dust particles, on average, to spiral in towards the Sun), and by the outward sweep of the pressure of sunlight.

While the dust particles providing the zodiacal light are in all parts of the sky, the glow historically studied by astronomers is found in a zone centered on the ecliptic where the greatest densities of particles are found. The light scattering properties of the dust combined with their maximum density near the Sun provide the glow commonly referred to as the zodiacal light, but more correctly called the zodiacal light pyramids.

Typically the tepee-shaped *zodiacal light pyramid* stretches to an elongation of 70 to 90 degrees from the Sun. However, the glow does not end there. Opposite the Sun is a fainter spot of light known as the *gegenschein*, a large oval patch of light stretching about 15 by 10 degrees along the ecliptic and centered at 180 degrees from the Sun. Joining these two is the extremely faint *zodiacal band* [Fig. 6.3].

154

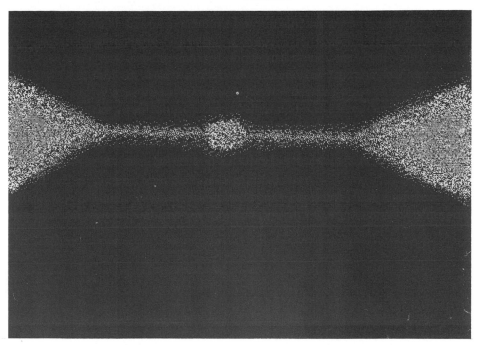

Figure 6.3   Schematic diagram of the zodiacal light, with the zodiacal band connecting the pyramid on one side of the sky to the gegenschein and continuing on to the other pyramid.

The pyramids are triangular-shaped glows seen in the western sky after the end of evening twilight and in the eastern sky before morning twilight [Fig. 6.4]. (The beginning and end of twilight are given in the *Astronomical Almanac, The Observer's Handbook* of the Royal Astronomical Society of Canada, and in the *Sky Gazer's Almanac* to be found in the January issue of *Sky and Telescope* magazine each year.) Although the pyramids may appear as bright as the Milky Way, they generally lack any texture. They may reach 45 degrees above the horizon and may cover as much as 45 degrees in azimuth, centered on the ecliptic–horizon intersection.

The zodiacal light pyramids have certain seasons when they are more easily seen. Specifically, you will have the best views when the ecliptic is as close to vertical as possible in the morning and evening. Thus, for the northern hemisphere, springtime is ideal for seeing the evening pyramid and autumn is best for the morning pyramid. The seasons are reversed for observers in the southern hemisphere, while observers in the tropics have good displays of both pyramids all year round.

The gegenschein, a German word meaning counterglow, is the very faint glow that appears directly opposite the Sun in the sky. It is centered 180 degrees away from the Sun. The gegenschein is caused by sunlight backscatt-

Figure 6.4   All-sky photograph showing the western zodiacal light pyramid on the right and the glow of Los Angeles in the southeast at lower left. The winter Milky Way crosses through the zenith. North is at the top. Note Orion just left of the meridian, below center. This original image was a 10 minute exposure with an 8 mm f/2.8 lens on Fujichrome 1600D. Photograph by M. Coco at the Polaris Observatory Association site in Lockwood Valley, California.

ered toward the Earth by interplanetary particles. Early theories of the origin of the gegenschein suggested that particles in a comet-like tail of the Earth or particles at the Earth–Sun Lagrangian L_3 point (found along the radius passing from the Sun through the Earth to the L_3 point) were the source of this counterglow. Studies by the Pioneer 10 and 11 spacecraft have shown, instead, the inner solar system is suffused with particles which scatter sunlight and make the zodiacal light.

Although it covers a large area of sky in the darkest region of the sky, the gegenschein cannot be seen from suburban locations. Seeing the gegenschein is a test not only of the observer's sensitivity, but also of the quality of the observing site's transparency and lack of sky illumination. It is not often observed from the darkest locations unless an observer especially looks out for it. The authors have tried to point out the gegenschein to novice observers, with varying degrees of success. A dark-adapted person who looks at the correct place, and with help from averted vision, should have no trouble detecting

Table 6.1. *Sighting the gegenschein*

Month	Constellations	Comments
January	Gemini–Cancer	In/near the Milky Way
February	Cancer–Leo	Near Regulus
March	Leo–Virgo	
April	Virgo	Near Spica
May	Libra–Scorpius	Near/in the Milky Way
June	Scorpius–Sagittarius	In the Milky Way
July	Sagittarius–Capricornus	In/near the Milky Way
August	Capricornus–Aquarius	
September	Aquarius–Pisces	
October	Pisces–Aries	
November	Aries–Taurus	Near the Pleiades
December	Taurus–Gemini	In the Milky Way

this large, faint patch which is slightly brighter than the background sky. Since it is centered at opposition, it is most easily seen around local midnight.

Other than the sky condition, two factors affect the visibility of the gegenschein – its elevation and its position relative to the Milky Way. It is harder to see in the summer seasons of both hemispheres, when the opposition point is low in the sky. The gegenschein is most easily seen during winter months around midnight when it is highest in the sky. It is virtually impossible to see it when the opposition point is covered by the Milky Way in Gemini, and completely impossible when it is covered by the Milky Way in Sagittarius. At high latitudes, summer observations of any component of the zodiacal light are difficult because of interference with the long twilight. Table 6.1 summarizes the visibility of the gegenschein.

The most difficult part of the zodiacal light complex to see is the zodiacal band. It is a barely perceptible belt of faint, hazy light that stretches from the end of the zodiacal light pyramid to the onset of the gegenschein [Fig. 6.5].

6.3 Visually observing the zodiacal light

While the quality of a night and an observing site can be evaluated by a look at the zodiacal light phenomena, other observations of these glows are less utilitarian but considerably more interesting observationally and scientifically. Binoculars and telescopes are useless for observing zodiacal light phenomena but an 'eye tunnel', a large diameter tube or open-ended box blackened on the inside, is useful in keeping extraneous, distracting sky or ground light out

Figure 6.5   All-sky photograph faintly showing the zodiacal band crossing the sky approximately horizontally. The glow of Los Angeles is in the southeast at lower left. The winter Milky Way crosses to the right of the zenith. North is at the top. Note Orion slightly to the right of the meridian, below center. The original image was a 20 minute exposure with an 8 mm f/2.8 lens on Fujichrome 1600D. Photograph by M. Coco from the Polaris Observatory Association site in Lockwood Valley, California.

of the observer's peripheral vision. The tunnel works like a horse's blinders when looked through. A sky crossbow [Fig. 6.6] or similar measuring device will be useful for visual angular measurements.

A sky crossbow can be made with a yardstick or meter stick attached at its middle to a wooden or plastic rod to make a T-shaped device. The center of the stick should be 57 units from your eye. For example, to use the inch markings on a yardstick for 1 degree intervals you should make the support rod 57 inches long, and hold the end of it to your cheek for measurements on the sky. For centimeters to correspond 1:1 to degrees, the support rod should be 57 cm long and if a meter stick is used it should be shortened to 36 cm. For better accuracy, use string to pull the yardstick/meter stick into an arc so that its ends are also 57 units from the end of the rod. (In fact, the meter stick can retain its full length as long as the ends are maintained 57 cm from the end of the support rod.) Table 6.2 lists easily recognized star pairs that can be used for checking the calibration, i.e., the length of the staff, of your sky crossbow.

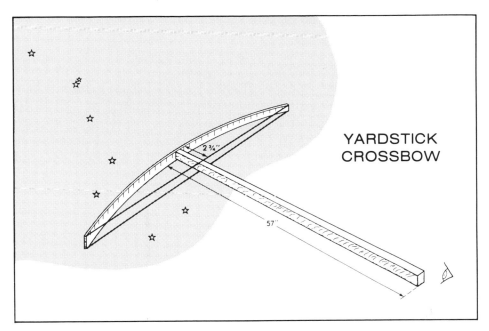

Figure 6.6   A sky crossbow. Measurements in centimeters (or other units) can directly replace values in inches if desired. (See text.) Courtesy of Sky and Telescope.

Table 6.2. *Sky calibration distances*

Star Pair		Separation [°]
α Boo	α Vir	32.8
α Boo	β Leo	35.3
α Boo	ζ UMa	37.1
α Boo	α Lyr	59.1
α Lyr	α Cyg	23.8
α Aql	α Lyr	34.2
α Aql	α Cyg	38.0
α Aql	α Sco	60.3
α Ori	α CMa	27.1
α Ori	α CMi	26.0
α Ori	α Tau	21.4
α Tau	α Aur	30.7
α Cen	α Cru	15.6
α Cen	α Car	58.0
α Cen	α Eri	61.3

Visual observations of the zodiacal light pyramid should include its maximum angular altitude and the width of its base. Also record the date and time of your observation and the limiting magnitude of the stars visible near the horizon, at 45 degrees altitude, and at the zenith.

Planets and bright stars, or groups of stars, near the ecliptic that are high in the sky at the end of evening or beginning of morning twilight can give a false impression of a boundary to the pyramid. Since the eye is attracted to the object, which is on the ecliptic in the same direction as the pyramid, the mind chooses it as the limit of the pyramid, when in fact it might extend several more degrees.

The triangle may not be symmetric because of the tilt of the ecliptic with respect to the horizon. A sketch of the limits of the pyramid and the position of the horizon on a star atlas serve to record the shape.

Once these basic observations have been made, check on the uniformity of the glow. Does it smoothly trail off from center (on the ecliptic?) to its soft edge? Is the triangle oddly shaped, with perhaps an arm sticking out from one side? Is any color detectable besides solar yellow-white? [Fig. 6.7]

It is common to see changes in the appearance of the pyramid. David Levy has observed it at times to consist of a double cone, a brighter outer pyramid surrounding an inner one that is fainter. Atmospheric seeing possibly has a considerable effect on the visibility of the light pyramid. David Levy has noted that some of the best displays are on nights when the seeing is very steady just before a storm. Observers have also reported a cyclical change in the pyramid that follows the sunspot cycle.

The observer scanning from horizon to pyramid top may find there is no apparent apex when conditions are very good: the zodiacal band is the extension and may be followed all the way across the sky. Stephen Edberg had this experience during a Perseid meteor observing session. As he followed the zodiacal light from the eastern horizon upward he found himself swinging over all the way to the Milky Way setting in the southwest! If the band is suspected, scan a part of the sky where you believe you see it perpendicular to the ecliptic and see if a band of increased luminosity is visible. Once the band is confirmed, make estimates of its width at several points. See if the band's axis is on the ecliptic or above or below it, or tilted across it by comparing it to a star atlas plot of the ecliptic.

If the time of night and conditions are right the zodiacal light may be traced from the pyramid along the band to the gegenschein. Even if the gegenschein is observed independently of one or both of the other zodiacal glows, measurements of its length and width are in order. Look at its shape and determine its center. Use an almanac later to see if the center is exactly opposite the Sun in the sky. If the center is not opposite the Sun it could be due to density variations of the particles, bias due to atmospheric effects, or perhaps to observer error.

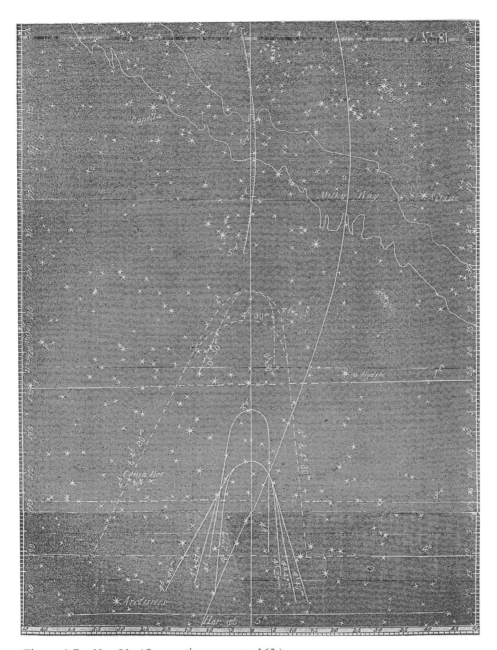

Figure 6.7   No. 81. (See caption on page 163.)

161

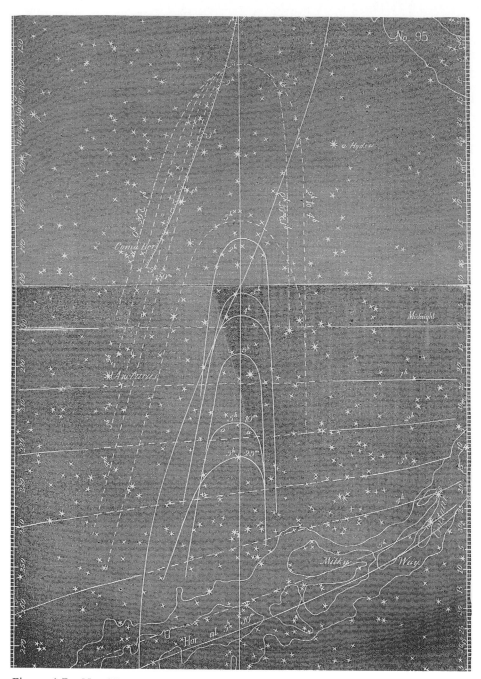

Figure 6.7   No. 95.

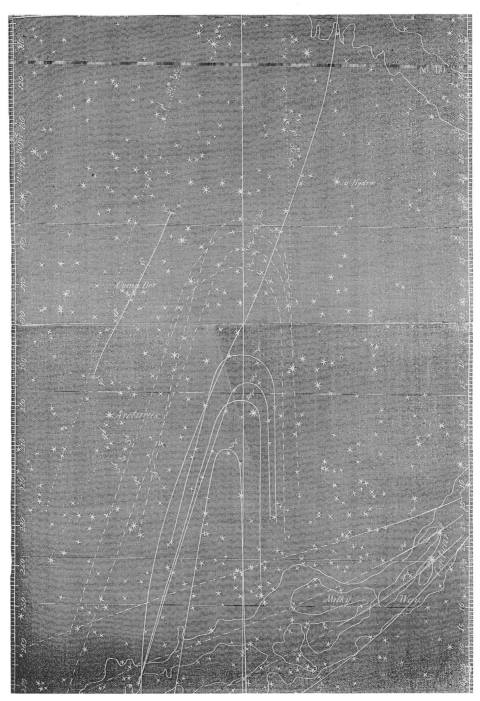

Figure 6.7 These pictures from Jones (1856) show some of the variations in position and shape that he observed in the zodiacal light pyramid in the course of a single night. He distinguished between a brighter light (solid line) and a dimmer light (dashed line). The horizon is indicated for times near those of the drawings of the pyramid. Observation No. 81 was made on 2 November 1853, No. 95 was made on 2 January 1854, and No. 110 was made on 30 January 1854.

Observations of the zodiacal light over a decade or two may show the effect of the sunspot cycle. The changes are not so much due to changes in the zodiacal light, but to changes in the airglow, a low level luminosity of the Earth's atmosphere (not auroral in nature) which varies with the solar cycle.

The ecliptic zone 36 degrees long and centered on the Sun is the most difficult one in which to study the zodiacal light. Because it is so close to the Sun it is never free of twilight and so the true shape and intensity of the pyramid are not well known. During a total solar eclipse you may have the opportunity to see this region of the sky in the dark. If you are dark adapted and can tear your eyes away from the spectacle of the corona, you can try to study this area in detail. Start looking for the pyramid as soon as totality commences, beginning the search away from the eclipsed Sun. The glow's intensity should increase as you sweep in toward the eclipse. It reaches maximum in the corona, where, in fact, the F or Fraunhofer corona is caused by dust particles scattering sunlight. Actually, there is no distinct end to the F corona; it just makes a smooth transition into the zodiacal light.

For observations like these the chances of success increase with excellent sky transparency and a long eclipse. The long eclipse is helpful not just in providing more viewing time but also by making the sky darker because the Moon's shadow cone's intersection with the Earth's surface is larger.

Stephen Edberg's experience with the long 11 July 1991 eclipse was not as ideal as the discussion above might suggest. This total eclipse was remarkable not only for its duration but also for the unexpected size and brilliance of the corona. The sky was considerably brighter than expected, probably because of the corona, and the layer of dust from the explosive eruption of the Philippine volcano Mount Pinatubo contributed to a decrease in atmospheric transmission and an increase in scattering that also probably made the sky brighter. A fainter, sunspot-minimum corona will make zodiacal light observations during an eclipse more effective, and perhaps an eclipse with the Sun low, resulting in a darker sky background, will improve the prospects of a search.

6.4 Photographing the zodiacal light

Photography of the zodiacal light is challenging. An equatorial mount is necessary, since the light shares the diurnal motion of the sky. Long exposures, in the 15 to 30 minute range with films having speeds of ISO 400 or higher, are necessary. Even then, the zodiacal light may not be as easy to photograph as to see. This results from the eye–brain combination that gives us our vision. It allows us to concentrate our attention and shut out distractions, something that an impersonal camera cannot do.

Lens focal lengths of 50 mm or shorter and focal ratios of $f/2$ or smaller (faster) are recommended. The lens should have minimal vignetting at the corners since this fall-off in illumination can mimic the true intensity variations in the zodiacal light. The vignetting of the lens can be tested by taking a picture of a clear blue, daylight sky and confirming that the film is evenly illuminated across the whole field of view. Since this will likely not be the case, it is worth the effort when time and conditions permit to take a second exposure immediately after the first, offsetting the aim point a few degrees in any direction so that the zodiacal light being recorded is in a different place in the field of view. This will confirm that general sky glow plus lens vignetting is not producing a false signature of the zodiacal light.

You can use the photographic methods described for separating the gas and dust tails of a comet to improve your images of the zodiacal light. Sandwich a zodiacal light negative with an equal density positive of your blue sky vignetting test. Printing the sandwich will subtract the effect of vignetting and show the true shape of the light, if it is done well.

6.5 Dust satellites of Earth?

Kazimierz Kordylewski began a search in 1956 for faint starlike objects at the Lagrangian L_4 and L_5 points in the Earth–Moon system. He reported success in 1961, describing visual sightings of faint clouds near both L_4 and L_5, and photographs of two clouds near the lunar L_5 point. (References for this and other reports mentioned below, written by anonymous editors, will be found by title in Appendix VI.) Observers were encouraged to try to verify these results. Hodgson (1962) reported little success but Simpson (1967) saw more. In 1975 J. R. Roach made observations with the satellite OSO 6 (Orbiting Solar Observatory) as the L_4 and L_5 positions were eclipsed in the Earth's shadow. He reported detections.

Almost 30 years after the initial report, another astronomer, M. Winiarski, has made a positive report. Using multiple photographs of the same sky areas when the L_5 point was in the field of view and when it would have orbited out, elaborate processing showed irregularities in the zodiacal band and cloud satellites being visible. The cloud satellites were several degrees across and wandered about the L_5 point. (Even Jupiter's Trojan asteroids do that, and it is not especially surprising given the Moon's elliptical orbit and the strong tidal influence of the Sun.) Winiarski claims the cloud satellites are redder than the gegenschein (but don't expect to see color visually) which could be the case if they have a different origin and composition than gegenschein materials, lunar for instance.

Can you see these extraordinarily difficult glows? Look for patches of light, one or two magnitudes fainter than the gegenschein, that are a few degrees across and positioned roughly 60 degrees ahead of or behind the Moon in its orbit. The *Astronomical Almanac* gives daily positions of the Moon; use Formula 3.1 in Section 3.5.2 to find the Moon's separation on the date of interest from its position 3 to 5 days ahead or behind that date. Look for one or more clouds near the past or future position that is about 60 degrees and a few days from the date of observation.

Observational conditions must be perfect, with a very dark and transparent sky, the Moon must be down, and the Lagrangian point of interest should be in opposition to the Sun so that backscatter will be maximized. (This implies the Moon must be in its gibbous phase, between first quarter and full or full and last quarter.) Avoid periods when the Lagrangian point is in the winter or summer Milky Way, and if possible, pick a time when the points will be well off the ecliptic so the cloud satellites will be distinguishable from the gegenschein and zodiacal band.

These observations will be a challenge, but additional confirmation with visual sightings or photographic or electronic images is desirable. Good luck!

7

Advanced observing techniques

7.1 Astrometry

The precise measurement of the position of a comet or asteroid with respect
to background stars, astrometry, is very important to orbit computations, ephemeris predictions, and, for comets, nucleus modelling efforts. By far the most
useful purpose of recording asteroids on film is to measure their positions
relative to nearby stars. An accurate ephemeris (list of predicted positions at
particular times) is an absolute necessity for the radio and radar astronomers
who may point at and track a comet or asteroid for long periods without
visually guiding on it. Because few professional astronomers are doing astrometry, the Central Bureau for Astronomical Telegrams and the Minor Planet
Center (same address, Appendix V) are seeking additional astrometric observations from amateurs.

Although the procedure is exacting, the results you get are extremely valuable because they can be used to provide a more accurate orbit. For newly
discovered asteroids and comets, astrometry is absolutely essential if a prelim-
inary orbit is to be determined quickly. In fact, the Minor Planet Center will not
credit a discovery of an asteroid until the discoverer has provided astrometric
positions for it over at least two nights.

7.1.1 Technique

A measurement accuracy of two seconds of arc or better is necessary. Measurements should be made with a measuring engine capable of measuring relative
positions to an accuracy of about 1 micrometer. These parameters suggest that
a camera focal length of greater than 200 mm will, in principle, give sufficient
image scale for accurate measurements. However, a longer focal length in the
range of 1000 mm to 2000 mm is strongly recommended. Penhallow (1978)
discusses the design, construction, and use of an astrometric reflector. However, a telescope need not be a dedicated astrometric instrument to provide
valuable astrometric data.

Astrometric photography should use the same technique as general cometary photography. The standard methods of emulsion development after exposure and, if desired, hypersensitization before exposure can be used.

Electronic images can be easily measured at the computer. The difficulty here is in selecting a suitable telescope focal length that provides enough position reference stars to fit on the small CCD.

We anticipate that more electronic hardware and analysis software will be available for astrometric reduction. This will considerably reduce the effort necessary for this work.

Guiding methods (1) or (2) described in Section 3.6.2 for cometary (and asteroidal) photography are the best to use when making astrometric photographs. Make the exposures just long enough to show the central condensation of the coma and some reference stars. Such an unspectacular comet photograph is the most easily measured since the stars are less trailed and the center of the condensation is more easily seen. The beginning and end times of the exposure should be recorded to the nearest second using Universal Time. Record details of the telescope used, observing site location (latitude, longitude, and altitude), and sky and weather conditions as well.

7.1.2 Measuring positions

Measurements should be made with an accurate measuring engine, a solidly-built machine which carries a photographic negative. The engine has a special negative tray that can be moved around in two directions ('x' and 'y') by turning very accurate lead screws attached to finely divided position indicators. Several stars and the target on the negative are centered individually under a microscope and their x and y positions are measured and recorded.

In the event your engine has only one screw, measure in one direction, rotate the negative 90 degrees, measure again, rotate 90 degrees once again and measure, and make one more 90 degree rotation for a final measurement. This procedure includes the re-measurement discussed below.

Measuring engines can be home-built (Penhallow, 1978, Everhart, 1982, and Balbi, 1987) if desired. Local observatories or universities may have a measuring engine that could be made available for astrometric measurements. No matter what measuring engine is used, it must be free of periodic errors in its screws. Even excessive oil on the screws can cause problems. It should also be as free of play as possible; to eliminate the effect of play, always approach the object to be measured from the same direction. Never backtrack if you overshoot.

Experienced users of measuring engines have developed 'tricks' to improve the accuracy of their measurements. Original negatives, emulsion side up, are always measured. Contact prints can be satisfactorily measured, but enlargements should never be used because of distortions introduced by the enlarger's lens. If the negatives are on film (and not glass plates), they should be sandwiched between two pieces of plate glass during measurement. Just as you sandwich the film with glass, you might want to cover the emulsion of the photographic plate with glass and draw a circle around each star you choose to measure, and then join the circles with lines to create a 'road map' that you can follow under the high power of the micrometer eyepiece. It may be easier to consistently have increasing x in the easterly direction and increasing y in the northerly direction.

At least three stars – ideally, eight or more stars – and the object must be measured on the negative. The stars should be evenly distributed all around the object, not situated on one side. A star whose position is known can also be measured as a check on the procedure. Faint stars with small images are easier to measure than bright stars with large images. Be aware that diffraction spikes may not be centered on a star's image. Measure the object during two stages of your process, once near the beginning, and again near the end. When setting on the object or star, don't ponder too long: set directly on it. Remove your hand from the screw between measurements of the same image so it won't 'remember' the position and unconsciously return to the same place.

The negative should be measured in one orientation and then rotated 180 degrees and re-measured to cancel out several kinds of error. Take the average of the two sets to use in your reduction. Finally, as suggested before, all images should be measured with the measuring engine screw traveling in the same direction. Do not remove the negative from the engine until all measurement is complete.

After reducing your data, you may eliminate the data from some stars whose positions are poor, but your final measurement should be accompanied by at least five stars if at all possible. If fewer stars are used this should be specified.

After you have recorded the x and y coordinates of a star, go back and do the measurement a second, and then a third time if the numbers in either axis are discordant. The position value you record for use should be the average of the three.

7.1.3 Data reduction

An astrometry reduction program is provided in Appendix IV, courtesy of Jordan D. Marché (1990) and *Sky and Telescope* magazine. It is complete enough to be useful as a good tool for reduction as you get started in this interesting and important field.

The mathematical reduction method that follows is designed for three stars and is the one suggested by Marsden (1982) and Marsden and Roemer (1982). Tatum (1982a) gives a very complete and readable discussion of comet astrometry. For improved accuracy, at least eight evenly distributed reference stars should be measured with the object. The method of least squares (see, for example, the chapter by F. Schmeidler in Roth, 1975) should be used to obtain a solution for the object's position. We emphasize again that you should really start this process with at least eight stars, expecting to finish with at least five because of inaccuracies in star catalogs. Use the following as a simplified example.

This procedure should be followed for each reference star [Fig. 7.1]:

(1) Identify a reference star on the negative and find it in one of the modern equinox J2000.0 catalogs such as the *Hubble Space Telescope Guide Star Catalog* (Lasker *et al.*, 1988), the *Position and Proper Motion Catalogue* (Röser and Bastian, 1991 and 1993), or similar catalog with J2000.0 coordinates for stars. Note the distinction between epoch, which is when the reference star catalogue observations were made, and equinox, which specifies a generally agreed epoch to which all coordinates and conversions are related. In many catalogs, epoch and equinox have the same value but they are not necessarily so. Check carefully before making computations.

Using the values of proper motion in right ascension and declination listed, update the right ascension (α) and declination (δ) of the star from 2000.0 to the date that the negative was taken. Do not precess the positions of stars or object to the current epoch, however. All results are referred to the 2000.0 equinox.

(2) Adopt a value for the position on the sky of the center of the negative in right ascension (A) and declination (D). Great accuracy in these values is not necessary. The same A and D for the center should be used for each reference star and the object.

(3) Compute the following values:

$$H = \sin\delta \, \sin D + \cos\delta \, \cos D \, \cos(\alpha - A) \tag{7.1}$$

As a check on H, note that its value should be approximately 1.

$$\xi = \frac{\cos\delta \, \sin(\alpha - A)}{H} \tag{7.2}$$

$$\eta = \frac{\sin\delta \, \cos D - \cos\delta \, \sin D \, \cos(\alpha - A)}{H} \tag{7.3}$$

Figure 7.1   Quantities used in measuring an object's position on a photograph are shown schematically here. The x, y coordinates of the target and reference stars are perhaps expressed in millimeters or inches. They are connected by equations to dimensionless 'standard coordinates' ξ, η on the sky. The x, y origin might be near the corner of the negative (it matters not where), but ξ, η are each zero at the right ascension and declination of the photograph's adopted center. Courtesy of Sky and Telescope.

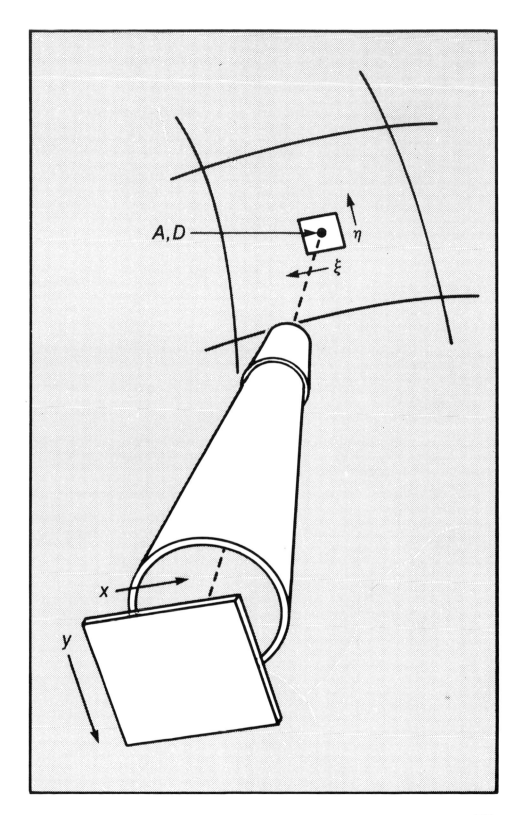

(4) Set the focus of the measuring engine's microscope on the comet's image and don't change it thereafter. Measure the star coordinates x and y on the negative with the measuring engine. If the stars are distinctly streaked from compensating for the target's motion, the ends of the trails should be measured and the average value used for the stars' positions at mid-exposure.

    The zero point for measurements on the negative can be anywhere on the negative but must be the same point for all measurements of the reference stars and comet. The units of x and y can be millimeters, inches, or whatever is convenient. Remember that all star and comet measurements should be made with the measuring engine screw traveling in the same direction.

(5) Adopt a value for the focal length (F) of the telescope and express it in the same units used for the quantities x and y. The value of F does not need to be known with great accuracy, but the same value should be used for each reference star.

(6) Generate the following equations:

$$\xi - \frac{x}{F} = ax + by + c \tag{7.4}$$

$$\eta - \frac{y}{F} = a'x + b'y + c' \tag{7.5}$$

where a, b, c, a', b', c' are unknown plate constants.

Repeating steps (1) through (6) for three reference stars will generate three pairs of equations (7.4) and (7.5), i.e., six equations for the six unknown plate constants. Using standard algebraic techniques commonly found in algebra books, the values of the plate constants may be found. As a rough check it should be found that a is approximately equal to b' ($a \approx b'$) and that b is approximately equal to $-a'$ ($b \approx -a'$). Using more stars over-determines these constants, allowing multiple-checks on the process.

Now measure the position x'' and y'' of the object on the negative. Using x'' and y'', the computed values of a, b, c, a', b', and c' and the adopted value of F, solve the equations:

$$\xi'' = \frac{x''}{F} + ax'' + by'' + c \tag{7.6}$$

$$\eta'' = \frac{y''}{F} + a'x'' + b'y'' + c' \tag{7.7}$$

The object's rectangular coordinates on the sky are ξ'' and η''.

Now solve the following equations in the order given to find the J2000.0 astrometric coordinates of the comet. The coordinates A and D adopted earlier for the plate center are used again:

$$\Delta = \cos D - \eta'' \sin D \qquad (7.8)$$

$$\Gamma = \sqrt{\left(\xi''^2 + \Delta^2 \right)} \qquad (7.9)$$

$$\alpha'' = A + \tan^{-1} \frac{\xi''}{\Delta} \qquad (7.10)$$

$$\delta'' = \tan^{-1} \frac{\sin D + \eta'' \cos D}{\Gamma} \qquad (7.11)$$

No attempt should be made to correct measurements for the effect of parallax. You can check for consistency by treating a known star as unknown.

Precise astrometric positions can only be achieved by experienced and disciplined observers. If you wish to make contributions in this area you should begin your observing program by refining your technique. There are only a very few amateur and professional astronomers who regularly contribute astrometric positions of comets to the Minor Planet Center. This is an under-represented observing discipline well suited to serious amateurs who enjoy the challenge of doing precision work.

7.1.4 Triangulation

With calculators and computers commonly available it is now easy for anyone to use surveyors' trigonometric expressions to triangulate the height and distance of a meteor. The technique described here can also be used by visual observers who plot meteors. However their results will generally be much less accurate because of inevitable perception differences and plotting inaccuracies.

Photographs of the same meteor from at least two sites anywhere from 40 to 100 km (25 to 60 miles) apart work best for triangulation. Although sites from 16 to 500 km (10 to 300 miles) are useable, short baselines make small, difficult-to-measure angles and there is some inaccuracy at the longer baseline distances because of the curvature of the Earth. The camera lenses, film sensitivities, and observing circumstances should be as similar as possible to ensure that the starting and ending points on the emulsions correspond to the same actual points on the meteor's path. The cameras should be aimed towards each other in azimuth and at an angular altitude equal to

$$Alt = 90 - (\tan^{-1}(\text{camera separation in km}/200)) \qquad (7.12a)$$

or

$$Alt = 90 - (\tan^{-1}(\text{camera separation in miles}/120)) \qquad (7.12b)$$

Recall that 1 km = 0.6215 mile.

The key data necessary for triangulation are the angular altitude and azimuth of the meteor from both sites. It is easiest to select the start, end, or any bursts in the meteor trail. In fact, measurements of the heights and positions of all of these events provide most interesting results. Note that using the start of the trail is somewhat unreliable because different angular distances from the radiant affect the apparent rate of motion and hence the likelihood that the meteor's point of commencement is the same for the two cameras.

Clock driven or stationary cameras may be used to obtain triangulation photo pairs, but even the use of leveled camera supports calibrated for altitude and azimuth will require knowledge of the camera's orientation, the scale of the photograph, and measurements of the negative.

When a meteor is believed to have been captured on film, the time should be noted to the nearest second and the camera's exposure stopped immediately, especially if it is on a stationary mount, to prevent some later misfortune from ruining the exposure. After the film has been developed, the star field around the meteor should be identified and the right ascension (α) and declination (δ) of the meteor's start, end, and any bursts should be determined for both negatives. This is easy with photos made tracking the sky; with negatives from stationary cameras the ends of the star trails are used to determine the αs and δs (hence the need to end the exposure immediately after the meteor). Plotting the meteor on a star chart may make the determination of the αs and δs easier. To identify the beginnings of the star trails some photographers put a 15 second break in them by blocking the lens 15 seconds to one minute after commencing the exposure.

The following computations in steps (1), (2), (3), (5), and (6) should be made for each start, end, and burst on each negative. The usual algebraic rules should be used when computing the expressions.

(1) Find the hour angle of the meteor:

$$G = T - \alpha \tag{7.13}$$

T is the local sidereal time, and may be determined using a commercial computer program, a graphic ephemeris like the *Sky-Gazer's Almanac*, the tables and formulae in the *Astronomical Almanac* and similar publications, or books on navigation or astronomical computation (like Fox, 1982). G should be converted to degrees: multiply the number of hours by 15, the number of minutes by 1/4, and the number of seconds by 1/240, and sum the results.

(2) Compute the cosine of the zenith distance and determine the angular altitude from that:

$$\cos Z = \sin \delta \ \sin P + \cos \delta \ \cos P \ \cos G \tag{7.14}$$

and

$$L = 90 - Z \tag{7.15}$$

where P is the latitude of the site.

(3) Calculate the azimuth U from

$$\cos U = (\sin \delta - \cos Z \ \sin P)/(\sin Z \ \cos P) \tag{7.16}$$

(4) From the latitudes and longitudes of the sites or from their positions on a map, determine the separation S of the sites and the azimuth R of the line between them, from the western to the eastern one. (The formula in step (8) to determine E may be used by substituting the sites' latitudes for the δ values and the sites' longitudes for the α values.)

(5) For the sites, referred to by the subscripts 1 (western) and 2 (eastern), determine the following values for the start, burst, and end:

$$D_1 = S \sin (U_2 - R)/\sin (2R + U_1 - U_2) \tag{7.17}$$

$$D_2 = S \sin (180 - R - U_1)/\sin (2R + U_1 - U_2) \tag{7.18}$$

(6) The height is

$$H_1 = D_1 \tan L_1 \tag{7.19}$$

which should equal

$$H_2 = D_2 \tan L_2 \tag{7.20}$$

as a check value. Sidgwick (1982) discusses correction factors for distant meteors and more advanced methods of triangulation.

(7) To find the path length begin by computing for both the start and end

$$r_1 = D_1 /\cos L_1 \tag{7.21}$$

and

$$r_2 = D_2 /\cos L_2 \tag{7.22}$$

(8) Using the linear trail length measured from the photo and the known image scale to get the angular length E, or by computing E using α and δ from either site with

$$\cos E = \sin \delta_s \ \sin \delta_e + \cos \delta_s \ \cos \delta_e \ \cos (\alpha_s - \alpha_e) \tag{7.23}$$

the actual path length N in the sky is

$$N = \sqrt{(r_{1s}^2 + r_{1e}^2 - 2 \ r_{1s} \ r_{1e} \ \cos E)} \tag{7.24}$$

where s and e refer to the start and end of the path. N may be checked by substituting r_{2s} and r_{2e} for r_{1s} and r_{1e} respectively; the final result should be the same.

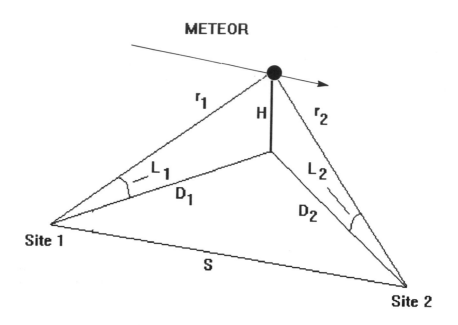

METEOR

r_1 H r_2

L_1 L_2

D_1 D_2

Site 1

S

Site 2

Looking Down onto the Earth's Surface

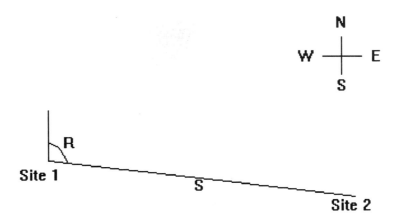

N

W ──┼── E

S

R

Site 1

S

Site 2

Looking Horizontally,
From the Side

To Zenith

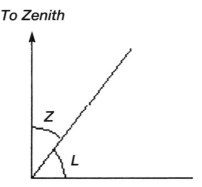

Figure 7.2   A meteor has a burst north of two observing sites. Some of the parameters necessary for computing the height of the event are presented in these three illustrations and defined in the text. Similar drawings can be made for the start and end of the meteor so that its path through the atmosphere can be computed.

Summary of symbols in the computations and in Fig. 7.2:
   α = right ascension
   D = ground distance from site to where event was seen at the zenith
   δ = declination
   E = angular path length
   e = end
   G = hour angle
   H = height
   L = angular altitude
   N = actual path length
   P = latitude
   R = site line azimuth
   r = line-of-sight range
   S = site separation
   s = start
   T = local sidereal time
   U = azimuth
   Z = zenith distance

While the computations for a triangulation solution are involved, they can easily be programmed for a calculator or computer. The results of such a computation are not only useful and interesting, but the process of computing them is also quite satisfying. (Do not be surprised, though, if you do not get values like H_1 and H_2 to match exactly. Measurement and round-off errors will cause discrepancies.) If the velocity of the meteor is deduced using a chopper, a rough orbit for the meteor can be determined. Using these procedures, Japanese amateur astronomers have discovered 41 new minor showers, complete with orbits.

7.1.4.1 An example of meteor triangulation

We offer here a worked example of meteor triangulation. It illustrates not only the process but also the difficulties involved in making these measurements, and the need to keep good records. Figure 7.3 shows a meteor captured from two sites by (1) R. Meier and E. Clinton, and (2) David Levy on 3 January 1981. Notice the small change in position of the meteor relative to the stars and the difference in sensitivity represented on the two pictures: the limiting magnitudes on them are quite different and so is the length of the meteor trail.

We cannot count on the starts and ends of the trails to be the same because of the difference in sensitivity (i.e., each camera probably started recording and losing the meteor at different times). Since there is no burst to use, we will assume that the meteor was brightest at the same point, in the middle, of its path.

The trails on each photograph have been carefully sketched onto a detailed star atlas [Fig. 7.4] and the midpoints are marked. Table 7.1 lists values used and derived in the course of our calculation. (Note that we carry four decimal places here, to minimize round-off errors in the calculation. The actual measurement precision is *not* this good.)

Unfortunately, records could not be found which recorded the exact time of the meteor or the geographic position of Site 1. We adopt values based on memory and a recent, approximate measurement of the position of Meier's site.

H has been rounded off to indicate less precision in the measurement. Not surprisingly, the values from the two sites are different, most likely due to measurement, site position, and time errors. The values are below typical meteor altitudes. A longer baseline and more accurate background information would improve these results.

7.2 Spectroscopy

There are three spectroscopic methods which amateurs can easily use with color or black and white emulsions or CCDs to obtain cometary or meteor spectra (singular: spectrum). The same techniques will work for minor planets but asteroids exhibit a reflected solar spectrum in visual wavelengths. It is only in the infrared that asteroidal spectra show interesting compositional details.

178

Figure 7.3 These photographs captured the same Quadrantid meteor in Leo in 1981. They can be used to determine the height of the meteor (see text). Photographs by R. Meier, E. Clinton, and D. Levy.

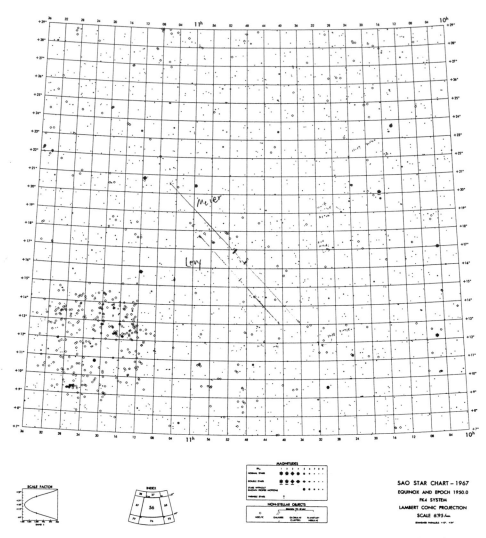

Figure 7.4   The images of the Quadrantid meteor in Fig. 7.3 were transferred to a copy of the Smithsonian Astrophysical Observatory Atlas (1969), *and then measured in preparation for triangulating the height of the meteor.*

Spectra reveal the principal chemical components of a comet or meteor. Dusty comets will be distinguishable from gassy ones, and all will generally exhibit the presence of molecules and radicals including C_2, C_3, CN, CO^+, H_2O^+, NH_2, and others. The size of the coma as seen in the various spectral lines is a direct measure of the excitation processes going on there. Meteors will show telltale spectral lines including iron, magnesium, calcium, and others.

Table 7.1. *Meteor altitude computation 3 January 1981*

	Meier and Clinton—Site 1	Levy—Site 2
Meteor	$\alpha=10^h50^m49^s=162.7042°$	$\alpha=10^h49^m34^s=162.3917°$
Midpoint	$\delta=16°55'=16.9167°$	$\delta=15°22'=15.3667°$
Site	$\lambda=110°46'16''$ west	$\lambda=110°46'$ west
Position	$\phi=31°55'36''$ north	$\phi=31°55'$ north
	$=31.9267°$	$=31.9500°$

These positions yield $S=2.6$km and $R=9.0°$.
For an adopted event time of 12:40 UT the local sidereal time is

T	12:09:07 = 182.2809°	12:09:08 = 182.2854°

The hour angle of the meteor is

G	19.5767°	19.8937°

The zenith distance of the meteor is

Z	23.2155°	24.5524°

The angular altitude of the meteor is

L	66.7845°	65.4476°

The azimuth of the meteor is

U	125.5867°	127.8491°
	$D_1=34.6182$km	$D_2=31.6233$km

The height of the meteor is

H	80.7100≈80km	69.2232≈70km

Note: A computation of the path length would yield wildly different values due to the difference between the sensitivities of the two cameras.

7.2.1 Methods and equipment

'Objective' spectroscopic observations require a disperser (prism or diffraction grating) to break down incoming light into its component wavelengths and a camera lens to focus the spectrum on the detector. 'Nonobjective' spectroscopy requires a disperser, detector, and telescope. A slitless spectrograph can be assembled which mimics observatory spectrographs in most of its features. No slits are necessary for any of these methods, though only the first one has a wide-enough field to easily capture meteor spectra.

For comets, the spectrograph should be guided on the comet using one of the methods described in *Tracking a Comet* (Section 3.6.2). The spectrum should have its dispersion perpendicular to the comet's tail. Virtually any size of camera or telescope objective can be used for these low dispersion spectroscopy methods.

For electronic imaging, camera/telescope focal length selection is important. The dispersion and placement of the disperser is also critical, since the spectrum must fit on the detector. A thin wedge prism or low dispersion diffraction grating (a Ronchi grating, for example) is necessary. Spectral analysis at the computer console is fun and interesting.

The choice of disperser will depend on a number of factors. Prisms are easier to find on the market but the results of their use may be harder to analyze. Diffraction gratings are harder to find but their data are easier to analyze. Ultimately, your choice may be based on cost and availability.

Prisms put all the light into one spectrum. Their disadvantages are that they absorb some light and do not disperse (spread) the light linearly. This means spectral lines are more crowded at the red end of the spectrum than at the blue end. Also, extra care is necessary to aim at the object of interest since they refract the rays as well as disperse them. A separate finder telescope or pointer is extremely useful for lining up on a comet (the unpredictability of meteors makes this accessory unnecessary for meteor spectra).

Diffraction gratings have a series of parallel grooves ruled on them, several hundred to the millimeter, which act to spread the light into a spectrum. They come in reflection and transmission varieties, and the best ones are very efficient. Dispersion depends on the number of grooves/mm and is nearly linear as a function of wavelength. Reflection gratings are hard to point at the object of interest, and a separate finder/pointer is almost a necessity.

With transmission gratings, the object and its spectrum (if it's bright enough) can be seen in the camera's finder.[1] The recorded spectrum is easier to analyze if a spectrum and its source are recorded together.

The disadvantages of gratings are that the grooves are delicate and easily damaged and that they produce a large (actually infinite) number of individual spectra, called orders, which make any particular order less intense than if all the light were going into a single spectrum.

A solution to this problem is found by adjusting the groove shape. Most of the light can be directed into one order. This is called 'blazing' the grating. Orders are numbered increasing outwards on both sides from the zero order (which is the direct, undispersed image of the object). Spectra generally get fainter with increasing order number except for the blazed order which is brightest by design. Orders starting with the second order and increasing have greater and greater overlaps on each other [Fig. 7.5].

---

[1] Author Edberg's experience shows that if the spectrum of a first magnitude star is visible when the star is viewed through a transmission grating with the naked eye, the grating is efficient enough for use in spectroscopic observations. Unfortunately, the cheapest gratings made of plastic film often do not meet this criterion.

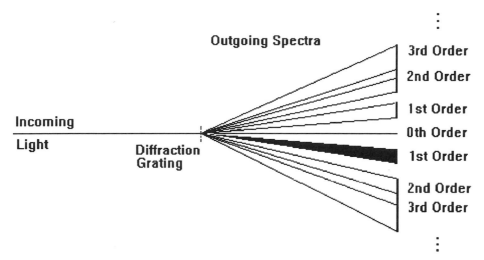

Figure 7.5   Light dispersed by a diffraction grating is split into many spectral orders. The 0th order is not dispersed but each order of greater numerical value has increased dispersion, and also overlap. As illustrated, the 3rd order overlaps on the 2nd, and the 4th (not illustrated) overlaps on the 3rd, and so on. In this schematic drawing, the grating has been blazed so one of the 1st order spectra is emphasized.

7.2.2 Technique

Objective prism spectroscopy by amateurs is discussed by Waber and McPherson (1967) and Patterson and Michaud (1980). Basically, a prism having a corner angle from 25 to 60 degrees is placed in front of a camera lens (50-mm to 200-mm focal length) so only light going through the prism reaches the lens. The prism and lens combination are chosen to ensure that the spectrum fits in the camera field-of-view. The prism should be oriented at the angle of minimum deviation, which means that the angle of incidence of light into the prism should equal the angle of emergence of light from the prism. The angles of incidence and emergence are measured with respect to the perpendiculars from the prism faces.[2]

To compute the wavelengths of spectral lines obtained with a prism, detailed information on the glass in the prism is required. It is easier to determine wavelengths using a calibration curve based on known wavelengths and their position on the film (Schmiedeck, 1979).

Mercury vapor lamps, unfortunately so ubiquitous, can provide useful wavelength calibration standards. The prominent hydrogen lines in A-type stars,

---

[2] To see this effect, look through a prism at an object and rotate the prism slowly on its axis. The object will appear to move, come to a halt, and move back in the opposite direction. The angle of minimum deviation is the orientation at which the image motion stops.

Table 7.2. *Selected spectral line wavelengths*

Source	Identification	Wavelength [nm]*
Mercury Lamp	Violet I	404.7
	Violet II	407.8
	Blue	435.8
	Green	546.1
	Yellow I	577.0
	Yellow II	579.1
A Star	Hydrogen (H)ε	397.0
	Hδ	410.2
	Hγ	434.0
	Hβ	486.1
	Hα	656.3
Comet—Coma	CN (0,0)	388.9
	C_2 (1,0)	473.7
	C_2 (0,0)	516.5
	C_2 (0,1)	563.5
—Tail	CO^+ (2,0)	427.3
	CO^+ (0,1)	550.2
	CO^+ (0,2)	624.3
	H_2O^+ (8,0)	619.8
	H_2O^+ (7,0)	654.4
Meteors	Calcium II (ionized) [K]	393.3
	Ca II [H]	396.8
	Sodium (Na) [D_2]	589.0
	Na [D_1]	589.6
	Magnesium (Mg)+Iron (Fe) [b_4]	516.7
	Mg [b_2]	517.3
	Mg [b_1]	518.4
	Fe—numerous lines, especially blue–green	

Note: Some lines may be blended at low spectral resolutions. Some molecular band heads are multiple, leading to differences in the wavelengths of the strongest head as reported by various authorities.
*1nm=nanometer=10^{-9} meter

such as Sirius and Vega, are also useful for calibration. The annual RASC *Observer's Handbook* and the *Astronomical Almanac* contain lists of bright stars and their spectral types if additional calibration stars are needed. Table 7.2 lists wavelengths useful for calibration and some found in various objects.

Objective and nonobjective grating spectrograph designs useful to the amateur (discussed by Edberg, 1982) work best if the grating is blazed for visual wavelengths (400 nm to 700 nm) in the first order. With objective grating spectroscopy, a grating with 300 to 600 grooves/mm is placed in front of a camera lens whose focal length is 35 mm to 100 mm (for standard 135 format film). Only light passing through the grating should reach the lens. The zero order and first order spectra should fit in the field of view. You may wish to obtain the second order in addition. The efficiency of the grating, film speed, focal ratio (*f*/number) of the optical system, sky brightness, and comet brightness play a part in determining the proper exposure. A minimum of five minutes on fast black and white emulsions or on hypersensitized fine grain emulsions is recommended.

Observers can identify the wavelengths of spectral lines by using the formula

$$\lambda = \frac{d}{n} \frac{L}{\sqrt{L^2 + F^2}} \qquad (7.25)$$

where

$$d = \frac{1}{\text{number of grooves /mm}}$$

n = order number of the spectrum used for measurement;

L = the distance on the film or plate from the zero order image to the spectral line in millimeters;

F = the focal length of the lens used in millimeters.

With nonobjective spectroscopy, the prism or grating is placed between any telescope objective (used as a light collector) and a plate or film holder (e.g. camera body) to hold the emulsion. A camera lens is unnecessary as the telescope acts to collect and focus light which is dispersed on the way to the film plane. It is important to focus on the spectrum and not on the source in this system since the disperser introduces coma, astigmatism, and field curvature.

A slow optical system is preferred because the smaller cone angle of the converging beam decreases wavelength uncertainty in the spectrum. The incidence angle falling on a disperser determines the position of a spectral feature. If that angle varies, as it does in a converging beam, the wavelength of the spectral feature is also uncertain in direct proportion to the uncertainty in the incidence angle.

This technique has been used with great success with telescopes as large as four meters for the detection of faint emission line sources such as quasars. When a grating is used, spectral lines may be identified according to the formula

$$\lambda = \frac{d}{n} \frac{L}{\sqrt{L^2 + D^2}} \qquad (7.26)$$

Figure 7.6   Nonobjective spectrum of Comet Austin on 28 April 1990. Many images of the coma, each in its own molecular spectral line, are visible in the first order spectrum, to the right of the zero order direct image. Numerous stellar spectra are also visible. Spectrogram by S. Edberg.

where D = the distance from grating surface to film plane in millimeters, and the other variables are defined above.

The head of the comet is the appropriate target for this method, and the dispersion should again be oriented perpendicular to the tail [Fig. 7.6]. The field-of-view is that of the telescope when used for direct photography. The larger the objective is, the better, since it makes this method more efficient.

For stellar sources, the focal ratio is unimportant, but it does matter for an extended source, in particular, the coma of a comet. When observing extended objects, a fast optical system is advantageous because a shorter exposure time is desirable. Thus, there is a trade-off between wavelength uncertainty and photographic speed, and instrument choice should be based on the goals of the observation when possible.

A slitless spectrograph can be constructed using a commercially available visual spectroscope (Lacroix, 1982). Light focused by the telescope objective is collimated by an eyepiece, passes through the direct vision prism system,[3] and is then focused by a standard camera lens attached to a camera body holding the film.

A slitless system has all the elements of an observatory spectrograph except

---

[3] Any prism or grating can be used. The direct vision prism system allows a simple, straight-through design without the bend in the optical axis that a single prism or grating usually requires.

186

the slit. As with objective prism spectroscopy, the determination of spectral line wavelengths is most easily accomplished with a calibration curve (Schmiedeck, 1979) when a prism or prism system is the disperser.

A slitless system may also be built using simple lenses with a disperser (Dahlmark, 1988). Obtain a pair of lenses of equal focal length, one positive and the other negative. (Optical supply houses and eyeglass dispensers are potential sources.) The lenses' focal lengths should match that of the telescope. The disperser should be sandwiched between the auxiliary lenses, with the negative lens on the objective side and the positive lens on the film plane side.

Remember that it is especially important to guide on the comet, rather than on a nearby star, when doing nonobjective or slitless spectroscopy. With any of the spectroscopic methods described in this section, poor guiding in the direction of dispersion will smear the spectral lines, making interpretation difficult or impossible.

7.2.3 Meteor spectrophotography

Meteor spectra can be photographed with a prism or diffraction grating placed in front of a fast camera lens [Fig. 7.7]. High-speed films should be used, and, as with direct meteor photography, hypersensitizing techniques should

Figure 7.7 A Geminid meteor spectrum captured at the edge of the field of view of a 40 mm f/1.7 lens, on Tri-X film, traverses the field diagonally. Stellar spectra extend horizontally. On the original negative, emission lines of iron can be identified as the group on the left. The bright line at far right is due to sodium, and the line between the iron group and sodium is due to magnesium. Spectrogram by S. Edberg.

improve the chances of catching a meteor. The camera should be firmly mounted on a stationary support (Majden, 1978).

Orient the camera so that the dispersion is perpendicular to the line from the aim point to the radiant (i.e. perpendicular to the path a shower meteor will follow). This aim point should be about 40 degrees from the radiant, although this is not a hard and fast rule. If desired, a shift in the aim of the camera and disperser combination in the direction parallel to the dispersion can be made to center the point in the field of view where the aim point's spectrum would be made. The film speed, lens focal ratio (f/number) and sky brightness determine the maximum allowable exposure for the field of view.

A lens shade is especially useful with a diffraction grating to prevent high order spectra of stray light from being imaged, but it can also eliminate useful high order meteor spectra should a fortuitously placed fireball occur outside the direct field of view. The choice of prism or diffraction grating should be based on the criteria discussed in the previous section but ultimately will probably depend on the availability of one disperser or the other.

The curious observer can identify the wavelengths of spectral lines by using exactly the same techniques discussed previously for comets. Libraries are good sources for lists of elemental wavelengths, as are some of the articles listed in the reference section of this book.

Millman and McKinley (in Middlehurst and Kuiper, 1963) list four classes of meteors grouped by their spectral characteristics. Note that $1 \text{ Å} = 10^{-10}$ meter $= 0.1$ nm.

> Type Y: The ionized calcium lines (Ca II) at wavelengths of 3933 Å and 3968 Å are the strongest lines in the blue-violet portion of the spectrum.
>
> Type X: If not as above, then either the sodium D lines (Na I) at 5890 Å and 5896 Å or the magnesium lines at 5180 Å or 3835 Å are the strongest lines in either the orange-green or the blue-violet portions of the spectrum, respectively.
>
> Type Z: If not as above, then spectral lines due to iron (Fe I) or chromium (Cr I) are strongest, in the orange-green or blue-violet.
>
> Type W: Not matching any of the characteristics listed in the other types.

Millman (1956), Russell (1959), Millman and McKinley in Middlehurst and Kuiper (1963), Harvey (1974), Feibelman (1974), Millman and Clifton (1979) and Schmidt (1984) contain additional photographs of meteor spectra.

7.3 Photoelectric photometry

Photoelectric photometry is an area in which a growing number of amateurs participate. The measurement of the brightness of objects using impersonal, electronic devices of high accuracy and precision generates very useful data for solving various astronomical puzzles. Accuracies of $+/-0.01$ magnitudes are possible and highly desirable.

The techniques of asteroidal and cometary photometry are not substantially different in principle from those of stellar photometry. Details of method are slightly different. Amateurs can now make observations in a new area that wasn't available to them even fifteen years ago. With comets, you can obtain interesting data because of the smaller image scale of your telescope compared to professional telescopes, if your equipment is sufficiently sensitive to avoid loss of low surface brightness data in equipment noise. This means a much larger area of the coma can be examined photometrically than can be done conveniently with large observatory telescopes. CCD detectors are now also being used, supplementing the photomultipliers used in the past.

There are several types of cometary observation to attempt. The determination of coma and central condensation surface brightness at specific wavelengths is the most obvious activity. When made with the proper filters, these observations lead to cometary gas and dust abundances, models of coma chemistry, data on the variation of cometary activity with heliocentric distance, and other results. Studies of short-period time variations are valuable. Coma and tail intensity profiles are of great interest if they are made with sufficient scale and sensitivity. Simple studies of the amount of polarization of light in the coma and tail can be made, but they are difficult to interpret. The photometry of a star seen through the coma or tail has only recently been accomplished.

Asteroids spin. Since these bodies are often irregular in shape, you can detect this rotation by simply monitoring the changing brightness, which can occur over a period of hours or days. Series of these measurements at several oppositions often permit determination of the direction of the asteroid's rotation axis. The standard B and V filters used for stellar photometry may be used.

However, we do not recommend attempting to determine the rotation period of an asteroid visually. Determining an asteroid's rotation is quite a challenge, since the variation is subtle, covering at most a few tenths of a magnitude over several hours, so that the bias of the observer could be greater than the full range of variation. Moreover, the asteroid's amplitude of variation varies depending on the aspect from which we view the asteroid. In trying to analyze visual data, we would not know whether a change was due to observer bias or to something real.

But if an asteroid will not give up its secret of rotation to an observer's eye, it will to the eye of a photoelectric photometer or CCD. With enthusiasm, diligence, good equipment, and lots of clear nights, you can unravel some of the mysteries of asteroids and participate in scientifically important projects helpful to the professional astronomical community: determining an asteroid's period of rotation and the orientation of its rotation axis.

The procedures are actually simple, although a photometer might seem complicated. If you are interested in developing a photometric program, here are some aspects you will need to consider.

The telescope

For asteroids, you are reading the light of a single point of light so the field of view you need is quite narrow; thus an $f/10$ to $f/16$ Cassegrainian system should work. The same is true for measurements of the central condensation of a comet, but a wider field of view will also yield interesting results.

Note that small Newtonian telescopes are awkward for photometry but have been used. The newer, lightweight photometers can be attached to typical Schmidt-Cassegrains and other telescopes that are widely available. The mounting of any telescope used for photometry must be very sturdy and accurately aligned on the celestial pole.

The photometer

A photometer is simply an instrument that measures light. As the photons pass through a telescope, they strike a detector which generates electrons in proportion to the number of photons striking it. Modern detectors called photomultipliers direct the electrons that come from their cathodes to other surfaces which multiply the current enough to be measured. Photodiodes, which are small, light-sensitive electronic chips, can also be used. Additional electronics are necessary to display the signal detected by the diode. A CCD imager with its array of light-sensitive cells on a single electronic chip can be used for photometry when it is followed by the appropriate electronics.

The results of photomultiplier and photodiode measurements are sent to a strip-chart recorder, digital meter, or computer. The strip-chart recorder consists of a revolving drum over which a strip of graph paper passes. The output from the photometer is continuously recorded, giving a complete picture over the night of the change in intensity of the light source. The photometer's output can be directed to a digital meter, whose output is recorded by hand, or better yet, to a computer which reads, records, and even interprets the counts fed to it.

A conventional photometer is useful only if all you want to do is measure light. It does not form an image of what you are measuring; it provides graphs on a strip-chart or a number. The beauty of what the photometer sees is all hidden in the numbers it provides, numbers which have to be interpreted before you can understand their meaning. A set of data that shows an asteroid getting brighter and then fainter does not necessarily mean that the object is really doing that. What if the comparison star that you have chosen also brightens and then fades, and then you determine that both objects reached their maximum brightnesses at the moment they crossed the meridian? Then all you have is a record of the changing apparent brightness of both the target and the comparison star as they rise and set and are seen through changing amounts of atmosphere. But if you subtract the effect of the rising and setting of the comparison star, which should not vary in actual brightness, the new result should represent the true variation of the object.

CCDs

There is another way to do photometry that lets you record images at the same time. This involves a charge-coupled device or CCD, a computer chip that looks very much like the chips in your home computer. The main difference between CCDs and computer chips is that CCDs are covered with an array of picture elements, or pixels, each of which is sensitive to light. A small CCD may have an array of 192 columns of 165 pixels each; larger ones used by professionals have as many as 2000 × 2000. Near the chip is a shutter which opens to allow light to strike these pixels.

While a typical photographic emulsion saves a small percentage of the light that actually strikes it, CCDs are far more efficient. They detect and record as much as 90 percent of the incident light. As soon as the shutter is closed, the chip transfers its data into a file in a computer, which then displays the image on a screen.

CCDs have an inherent problem in that their pixels do not all have the same efficiency. The process of flat-fielding is intended to solve this problem. If you point your telescope at a bright light reflected off the observatory dome, or at a twilight sky, the resulting image will show a rather strange Rorschach ink blot appearance. This is known as a flat-field. By performing an electronic ratio between it and the raw image, the result is a much cleaner image. Also, chips have electronic noise or 'bias' levels which must be subtracted from the raw image as part of this process of reduction.

How do you do photometry with a CCD system? Theoretically, the process is simple. Even though an asteroid is essentially a starlike point source of light, atmospheric currents always make it appear to take up a small area of sky. Comets also cover a small area of the sky, but usually larger than a stellar seeing disk. Your first step is to flat-field your image. Then, you outline an area around the most exposed pixel, large enough to include every pixel which has detected a photon from the object. Then you read another area that includes no starlight at all – nothing but the brightness of the sky background. By subtracting the second figure from the first, you have an actual reading of the object's light output. After processing the comparison star images in the same manner you can compare all these data.

7.3.1 Observing procedures in photometry

The reduction of a photometric data set is much easier if each step of the process of obtaining it has been taken with care and precision. The first concern, of course, is the weather. If there are any clouds at all in the sky, especially cirrus clouds, the object will appear to vary in a haphazard manner quite independent of its intrinsic variation. The comparison star will also vary in a random way as clouds pass by, with a net result of useless data. If you are obtaining your data with a CCD, you actually obtain an image of the field, complete with comparison stars taken at the same time as the asteroid. In this case, the effect of the clouds on the asteroid and comparison star is approximately, though not exactly, the same, and your data are useful if they can be compared with images of the same field you obtain on a completely cloudless night.

There is a difference between 'relative' and 'absolute' photometry. In the first instance you compare the magnitude of the asteroid with that of one or two comparison stars, the second star being known as a check star. These data should be consistent over a night, or over an observing run of several nights, and are useful if all you are after is a basic idea of the light curve of the asteroid. Absolute photometry, on the other hand, provides a much more complete picture of the asteroid's rotation over long periods of time, as the asteroid orbits the Sun over several years. With absolute photometry, conditions should be completely cloud-free all the time, a condition we call 'photometric'. The reason is that in absolute photometry, a single set of standard stars whose magnitudes have been precisely determined, and which are in different fields from that of the asteroid, needs to be observed as well and used as a baseline for all measurements.

7.3.1.1 Observing asteroids with a photoelectric photometer

In a program designed to obtain many light curves over a full orbital period for several asteroids, David Levy joined the Planetary Science Institute in the following procedure for observing with a photometer. Other observers use different procedures, but they are essentially variations on this theme. We began with the first comparison star, which we use for focusing and testing. When we were satisfied with the consistency of our results, we then measured the comparison star over three integrations, and then moved on to the asteroid.

We sampled each of our asteroids at least six times per hour in one color (V filter) with a target precision of 1 percent photometry (0.01 magnitude). Since we needed to make an observation every ten minutes, it was possible by working rapidly to interweave data points on two or even three asteroids, and thus obtain several light curves simultaneously.

Most of our data have been acquired at what was known as the No. 2 36-Inch Telescope (90-cm) at Kitt Peak National Observatory, an ideal telescope for our program. This telescope's large aperture allowed us to observe to magnitudes as faint as 15. More importantly, the telescope had high pointing accuracy and was computer controlled, allowing us to find targets quickly and efficiently.

Positional information for each asteroid and comparison star was stored in the telescope's computer during the day, to get called up as the telescope slewed across the sky toward the position of the first asteroid's comparison star. The photometer began a series of three 5-second integrations of the comparison star, after which the telescope automatically offset a minute of arc, usually to the north, so that it measured the sky background which was then subtracted. (On occasion we have also used a smaller 40-cm (16-inch) reflector, with which we would manually find and measure the asteroid, comparison star, and sky background.)

The same observing procedure was then followed for the asteroid, which was usually within a degree of the comparison star. Enough integrations allowed us to obtain at least 20 000 net photon counts, a number which provided the required 1 percent precision if the sky was dark and moonless. On bright and moonlit nights we needed more counts.

We repeated this procedure for up to three asteroids at once, and since our program required us to observe each asteroid six times per hour, the nights were busy. When we used the smaller telescope with its manual controls, we would be limited to one asteroid, comparison star, and standard set at a time.

As our observing night advanced, we continued to gather data at a frenetic pace. Moving the telescope from comparison star to background to asteroid to background to new object, we tried to return to the first object within 10 minutes, all the while keeping watch on the sky for possible cirrus. Because there is no easy access to the outside from our dome, the sky watch was a cumbersome but necessary adjunct to our observing program. As the data continued to come in, they were checked in two ways – first, the three or more integrations that make up a single observation should be within one percent of each other, and second, the comparison stars should remain consistent with earlier readings. At high air masses, when the telescope was pointing near the horizon, the comparison stars appeared to brighten or fade more radically. Although each asteroid was assigned but one comparison star, at any hour we observed two, and possibly three, comparison stars.

In addition, we observed a series of Landolt standard stars – objects for which accurate magnitudes had been determined – at six opportunities each night, in order to serve as controls on the atmospheric and other conditions of our program. For each observing run we chose three stars, whose characteristic magnitudes are well known, from the list of Selected Areas that was published by Arlo Landolt in 1973. The evening set was chosen so that it was at a low air mass (between 1.0, with telescope pointed near the zenith, and 1.5) within two hours after the evening began. The three stars were observed in two colors, and then their backgrounds were also observed, but since the three objects were within a degree of each other, finding them with the telescope was not as time consuming as it appears.

We next observed the standards when they were at about 1.8 to 1.9 air masses and then again between 2.3 and 2.5 air masses. This final reading occurred generally just before midnight, and shortly after midnight we began the same procedure with a new set of standards just rising in the east. We tried to adjust our asteroid reading procedure at this point. To reduce the time lost because of the need to read standard stars, we reversed the order and read, for one cycle after a standard set, the asteroid first and then the comparison star. Moreover, by reading the standard stars when the telescope was already near a target asteroid, we reduced the time lost between asteroid readings. The standards were important because they provided a constant check on the consistency of our data at various air masses.

7.3.1.2 Observing with a CCD

A CCD involves a different observing procedure. As with a photometer, it is important to center the object of interest precisely in the field of view, but unlike a photometer, it helps to find a portion of the chip that is free of any defective pixels that would interfere with the accuracy of the photometry you attempt. The experience that David Levy has with this sort of photometry was gained during a long series of nights during which he and Wieslaw Wiesniewski of the University of Arizona tried to obtain a light curve of Comet P/Halley. Since the comet was faint, we needed ten minute exposures to record the comet with sufficient strength, without the sky brightness rising as well. On moonlit nights, we needed to make shorter exposures, because once the sky brightness began to increase, exposing longer would not increase the value of the observation.

We would use stars in the same field of view for each night's comparison sequence, and would separately record images of three standard stars three times during each night's work.

7.3.2 Photometry of comets

The CCD studies of Halley's Comet described above were designed to answer questions regarding the nucleus of the comet. Specifically, they were intended to look at the photometric behavior of the nucleus (rotation rate, activity variation, etc.) and treated it as an asteroid. Studies of the coma are different.

Changes in hardware and in observational methods are necessary for the generation of usable photometric data on cometary comae. The most important hardware change is in the filters. The standard U, B, and V filters used for stellar photometry are virtually useless on comets because their large bandwidths do not allow separation of the contributions of the gas from those of dust in the light of the comet.[4] Professional astronomers use the standard filters

---

[4] A V magnitude is crudely convertible to a visual magnitude. Studies in the area of photoelectric vs. visual magnitudes are interesting to some researchers, but the value of such data is questionable. U, B, and V filters can be used for transits and occultations of stars by the comet (mentioned elsewhere in this section). These filters are, as mentioned before, very poor substitutes for a set of comet filters for making measurements of the comet.

Table 7.3. *Cometary photometry filters*

Species	Source	Central Wavelength (Å)	Bandwidth (Å)
OH	gas	3089	60
Blue Continuum	dust	3650	80
CN	gas	3871	50
C_3	gas	4060	70
CO^+/N_2^+	gas	4260	65
Mid-Continuum	dust	4845	65
C_2	gas	5139	90
H_2O^+	gas	7000	175
Red Continuum	dust	6840	90

Note: 1 Å$=10^{-10}$ m

listed in Table 7.3 as well as many other narrow passband interference filters for their photometric work on the coma. Some of those filters are available on the commercial market, and may even be found in manufacturers' surplus and overrun lists.

The diaphragm size and image scale are directly related. Most large professional telescopes have large image scales and use small diaphragms. An important amateur contribution can be made when a comet head is at its greatest angular extent. The small image scales of amateur telescopes combined with a large diaphragm allow much more of the head to be measured than is easily possible with professional telescopes, since photomultiplier tubes and CCDs have limited active surface areas. The diaphragm size in linear and angular measures must be determined with great accuracy for proper reduction of the data. Use an accurate caliper for the former and time the passage of a star across a diameter of the diaphragm for the latter (cf. Section 3.5.2, Formula 3.2).

Procedurally, the data are reduced in the usual manner, but more data must be obtained in a more careful manner. Extinction and sky brightness corrections must be made very carefully for each observation because the comet will often be observed low on the horizon around sunrise or sunset. Furthermore, comets are extended objects of very low surface brightness in their outer parts. Average extinction values for the amateur's observatory should never be used. Instead, be sure to make extinction measurements of the sky using standard stars every night comet photometry is done. (There are situations when there is just not enough time to make both the standard calibration measurements and measurements of the comet itself. In such circumstances it is better to just enjoy observing the comet some other way. The importance of the calibration data is such that without it the comet measurements are of little value.)

The books *Photoelectric Photometry of Variable Stars* by Hall and Genet (1981) and *Astronomical Photometry* by Henden and Kaitchuck (1982) are excellent guides to standard data reduction techniques. A'Hearn discusses comet photometry in Genet (1983b).

Photoelectric magnitudes of the comet should be measured in a manner similar to that for measuring stars. The sequence of measurement should be dark current, comparison star, comet and sky several times, comparison star, dark current, and then the sequence should be repeated. Each sequence should consist of observations with at least one molecule (gas) filter and one adjacent continuum filter. After several sequences with one diaphragm, the set of observations should be repeated with a diaphragm of another size. If there is any evidence of clouds or a change in atmospheric transparency (for example, a comparison star has a weaker signal even though it is higher in the sky) photometric observations should be suspended.

Standard stars are listed in Tables 7.4, 7.5A, 7.5B, and 7.6, copied from Edberg (1983) and the included citations. Additional standard stars, including their magnitudes in the cometary filter set, will be found in Osborn *et al.* (1990) and in the International Halley Watch CD ROM archive.

Sky brightness measurements should be made at least 1 degree away from the comet head and tail to avoid contamination by the faint outer coma (which will not be visible to the eye). This is crucial to the validity of the comet measurements. Dark current measurements are not necessary in each sequence if past experience has shown them to be stable with time, temperature, etc. (They are not needed in single photomultiplier tube systems if sky measurements are made carefully.) The most valuable results are possible only with a tube cooled to reduce the dark current. The time of each raw comet head measurement should be recorded to at least the nearest minute.

The observing sequence should be modified to include more comet measurements with identical settings if a significant change in apparent brightness is detected. Knots or disconnected pieces of the ion tail can also be followed through their evolution, but a CO^+ filter is necessary for such observations.

Time studies of intensity variations of the coma or central condensation (large or small diaphragm, respectively) are useful for studies of nucleus activity, rotation, and the comet's interaction with the Sun. Gas (the C_3 filter is recommended) or dust filters may be used, and observations should be made continuously through the night. Only one or two filters may be used during such observations.

In any circumstances, it is important to have the same portion of the comet in the diaphragm for each measurement and to record the location in some objective fashion. This requires care and skillful technique.

With a small diaphragm, intensity profiles across the comet can be obtained

Table 7.4. *Standard stars for IHW photometry*
Primary equatorial flux standards

Star identification

HD	Other	R.A.	1950	Dec.	V Mag.	B-V	Spectrum
		h m s		° ′ ″			
3379	53 Psc	0 23 47.7		+14 57 24	5.88	−0.15	B2.5IV
26912	μ Tau	4 12 49.0		+ 8 46 07	4.27	−0.07	B3IV
52266	BD −5° 1912	6 57 53.9		− 5 45 21	7.23	−0.01	09V
74280	η Hya	8 40 36.7		− 3 34 46	4.29	−0.20	B4V
89688	(RS) 23 Sex*	10 18 27.1		+ 2 32 32	6.68	−0.09	B2.5IV
120086	BD −1° 2858	13 44 44.2		− 2 11 40	7.89	−0.18	B3III
120315	η UMa	13 45 34.3		+49 33 44	1.86	−0.19	B3V
149363	BD −5° 4318	16 31 47.9		− 6 01 59	7.80	+0.01	BO.5III
164852	96 Her	18 00 14.7		+20 49 56	5.27	−0.09	BV3
191263	BD +10° 4189	20 06 15.1		+10 34 44	6.33	−0.14	B3IV
219188	BD +4° 4985	23 11 28.0		+ 4 43 29	6.9	−	BO.5III

Note: *Although it has a variable star designation, several more recent investigations find Δm < 0.01.

Table 7.5A. *Primary solar analogs*

HD	Other	R.A.	1950	Dec.	V Mag.	B-V	Spectrum†
		h m s		° ′ ″			
28099	Hyades−vB 64	4 23 47.7		+16 38 07	8.12	+0.66	G6V
29461	Hyades − vB 106	4 36 07.6		+14 00 29	7.96	+0.66	G5V
30246	Hyades − vB 142	4 43 38.9		+15 22 59	8.33	+0.67	G5V
44594	HR (YBS) 2290	6 18 47.1		−48 42 50	6.60	+0.66	G2V
105590*	BD −11° 3246	12 06 53.2		−11 34 36	6.56	+0.66	G2V
186427**	16 Cyg B	19 40 32.0		+50 24 03	6.20	+0.66	G5V

Note: †Spectral types are from various sources and apparently indicate differences in the classifiers rather than in the stars.
*Brightest member of a triple system; 8.9 and 9.1 companies at (1.6 E, 4″ N) and (0.5 W, 22′ S).
**Fainter member of a double system, 6. companion at (3. W, 28″ N).
Source: From Hardorp (1982 A&A **105**, 120 references therein).

Table 7.5B. *Other solar analogs*

Star identification

HD	Other	R.A.	1981	Dec.	V. Mag.
		h m		° ′	
—	SAO 120107	13 46.0		+ 6 07	9.26
120528	SAO 28894	13 48.0		+53 20	8.55
144873	SAO 65083	16 06.0		+34 09	8.84
—	+15° 3364	18 06.5		+15 57	8.64
191854	SAO 49262	20 09.6		+43 53	7.43

Table 7.6. *Faint equatorial solar analogs*

Landolt #*	V Mag.	Spectrum	R.A.	1975	Dec.
			h m		° ′
92-433	11.65	G2	0 55.6		+0 53
93-241	9.39	G2	1 54.0		+0 29
94-293	7.02	G5	2 54.0		+0 20
95-236	11.48	G2	3 54.9		+0 04
96-393	9.66	G1	4 51.2		−0 00
97-249	11.74	G2	5 55.8		+0 01
98-313	11.07	G0	6 51.5		−0 34
99-358	9.59	G3	7 52.7		−0 18
100-289	9.13	G7	8 52.5		−0 27
101-057	11.58	G3	9 56.8		−0 54
102-1081	9.91	gG2	10 55.8		−0 05
103-487	11.84	G3	11 53.9		−0 15
104-335	11.70	G3	12 41.1		−0 25
105-257	9.14	G0	13 37.1		−0 52
106-1146	9.10	dG2	14 42.5		+0 09
107-469	12.17	G0	15 38.4		−0 22
108-1911	8.04	G3	16 36.5		+0 06
109-381	11.71	G0	17 42.9		−0 20
110-529	11.41	G0	18 42.7		+0 25
111-1342	9.22	gK2	19 36.0		+0 13
112-636	9.85	G3	20 40.3		+0 11
113-459	12.13	G0	21 40.0		+0 36
114-651	10.28	G2	22 39.9		+0 55
115-366	12.11	dG3	23 43.0		+0 53

Note: *c.f. A.U. Landolt, 1973, AJ **78**, 959.

by allowing the comet to drift through the field of view. The drift should start at least 0.5 degree away from the comet and be at a smooth, constant, known rate (using either the Earth's rotation or the telescope axis drive motors), and the position of the path across the comet should be known. Beware of background stars included in the scan that can falsify the profile. This type of observation can be made in gas and dust filters on the comet. Extinction measurements should be made before and after the drift.

Polarization measurements of the coma are not generally of obvious value, although there are special cases of transient activity when they would be. They should be made using either a rotating polarizing filter or with several filters oriented in known, different directions (cf. Section 3.5.6). It is important to know the direction of polarization with respect to north (i.e., the position angle measured from north through east) of each measurement. Military surplus polarizers should be used, not the camera store variety. (Camera store polarizers are not as efficient as other polarizers.) Again, a clear record of the location on the comet for which the data were taken is absolutely necessary.

Photometry of a star as the coma or tail passes in front of it (a transit or, better yet, an occultation) would be very interesting. The normal techniques and filters of stellar photometry should be used, but brightness measurements without the star after the event will yield the appropriate intensity value for subtraction. Pure star readings and dark sky measurements should still be included at least one degree away from visible cometary features. Accurate timing of the event, both when it occurred and its duration, is critical. Such data have the potential to yield astrometric information and physical information including the size and density of the dust coma and the size of the nucleus.

7.3.3 Photometry of the zodiacal light

The photometrist looking for a challenge can try to detect and even map the zodiacal light. Detecting these areas of luminosity may be achieved by taking counts using wide field optics and a diaphragm limiting the field of view to be within the 'boundaries' of the zodiacal light patch, and then making an equivalent 'dark sky' observation well away from the patch. Excess counts are indicative of zodiacal light, though other sources including starlight and airglow can confuse the result. A truly proper test of detection involves measuring the zodiacal light and dark sky zones of interest when the zodiacal light is known to be minimal, and then a few months later when it is present in the target zone. This requires careful planning over perhaps several months but provides necessary calibrations and cross-checks to be certain of the result.

To map the zodiacal light the field of view should be a few arcminutes to a degree or so in diameter. The idea, as with the detection scheme discussed above, is to avoid as much as possible the detection of individual stars. The telescope and attached photometer should be scanned in right ascension and declination from well outside the luminosity boundary, through it, to beyond the other boundary. Depending on the sensitivity of the telescope and photometer system the measurements can be made with short integration periods, as the telescope is scanned at a constant rate continuously with a boustrophedonic (constant-rate, stepping zigzag, also called a box scan) pattern [Fig. 7.8] starting near the horizon and working upward in the case of the evening zodiacal light pyramid, downward for the morning pyramid. For photometric apparatus requiring long integrations the telescope is simply moved from site to site when the integration is completed.

Figure 7.8 A 'box' scan pattern, for measuring the extent of the zodiacal light. The measurement pattern extends well beyond the pyramid of light in altitude and azimuth to better delimit the illuminated area.

As with the detection scheme, careful repetition of the same aim points when the zodiacal light isn't present provides the best confirmation of the map. Careful record-keeping is important! For both detection schemes system noise (dark) counts must be made and accounted for.

A more complete approach to detection and mapping the zodiacal light should include measurements of airglow. Aim in a direction that is not contaminated by aurora, city sky glow, or the zodiacal light and make brightness measurements at various altitudes from the horizon upward, as high as your zodiacal light measurements are made. The airglow measurements will help calibrate your zodiacal light and empty sky measurements.

Photoelectric photometrists will find studies of the zodiacal light to be a challenging task. Long-term studies may reveal changes in the light, the observing environment, the solar cycles, and in the equipment used.

7.3.4 Conclusion

These projects give amateurs a challenging variety of ways to observe comets, asteroids, and the zodiacal light. Photometry is an involved process, leading to knowledge of fundamental characteristics of the object that can have practical value, for example, in planning spacecraft encounters with these fascinating bodies. A great deal of care must be taken to obtain usable photometry. Persistent and patient observers will be rewarded with new knowledge of these fascinating and mysterious objects.

Appendix I: Glossary

Altitude A measure of the angular position of an object, measured in degrees above the horizon; from 0 degrees at the horizon to 90 degrees at the zenith.

Amor asteroids Asteroids whose orbits permit them to come inside Mars' orbit.

Antitail Projection effects, when the Earth crosses the orbital plane of a comet, sometimes make a portion of the comet's tail appear to point towards the Sun.

Aphelion The farthest point in an object's orbit around the Sun.

Apollo asteroids Asteroids whose orbits permit them to come inside Earth's orbit.

Apparition The period of time that a celestial object is visible from Earth.

Appulse A near miss — the apparent close approach in the sky of two celestial objects. Physically they may be widely separated in space.

Astronomical unit, AU Defined to be the Earth's average distance from the Sun, 149.6 million km (93 million miles).

Aten asteroids Asteroids whose orbits are inside Earth's orbit. Their semi-major axes are less than 1 AU.

Azimuth A measure of the angular position of an object, measured in degrees around the horizon from the 0 degree point at due north, through 90 degrees at due east, 180 degrees at due south, 270 degrees at due West, and back to 360 degrees = 0 degrees at due north.

Bolide An exploding meteor.

CCD An abbreviation for charge-coupled device. A CCD is a highly efficient light-collecting silicon chip whose light-sensitive pixels (short for picture elements) record the amount of light that strikes them. The picture, or image, that the chip records is then read by a computer.

CHU Station broadcasting radio time signals for the Canadian government on frequencies of 3.330, 7.335, and 14.670 MHz.

Coma The extended area around the nucleus of the comet which has not yet been strongly affected by the solar wind or solar radiation pressure.

Commensurable, commensurability Used when discussing orbital periods having integral multipliers, e.g., two and four year periods (multiplier = 2) and two and six year periods (multiplier = 3) are commensurate, two and five year periods are not (multiplier = 2.5). An asteroid having an orbit commensurate with a large planet will suffer large gravitational perturbations until those perturbations change the orbit enough to reduce its commensurability with the planet's period.

Conjunction The alignment of two objects in the sky. Often used to imply a solar conjunction, when the object's visibility will be lost in the Sun's glare.

Dust tail Solid dust particles (blown off the nucleus of the comet as it sublimates), responding to solar radiation pressure and their orbital motion, are pushed away from the nucleus. The dust tail is seen because of sunlight scattered by the dust.

Ecliptic The path of the Sun in the sky projected on background stars.

Ephemeris A listing of predicted positions on the sky for a moving object. The International Astronomical Union, through its Central Bureau for Astronomical Telegrams, issues ephemerides or positions for every newly discovered comet and for many newly found asteroids. Through the Minor Planet Center in Cambridge, Massachusetts, astronomers can obtain ephemerides of any asteroid or comet.

Fireball A very bright meteor, generally defined as exceeding the magnitude of Venus (mag. −4).

Fluorescence The emission of light of a longer wavelength after absorption of shorter wavelength electromagnetic radiation.

Gegenschein Literally meaning 'counterglow,' this phenomenon of the zodiacal light refers to sunlight back-scattered from interplanetary dust located opposite the Sun in the sky.

Head (of a comet) Refers to the nucleus and coma together.

Hydrogen envelope (of a comet) Seen only in ultraviolet light, this gigantic cloud of atomic hydrogen surrounds the comet's head. Can be up to 10 million km in diameter.

Ion, ionize An ion is a neutral atom or molecule which acquires additional positive or negative electrical charge. Solar ultraviolet and x-radiation is the principal reason neutrals become ionized in comets. Charge transfer between ions in the solar wind and neutrals in comets can also cause ionization. High speed collisions can also ionize matter, as exhibited by meteors.

Ion tail The parent molecules released by the nucleus of a comet are ionized by sunlight and dragged away by the magnetic field carried by the solar wind to form the ion tail. The tail is seen by the light of fluorescing ions.

Kirkwood gaps   Areas in the asteroid belt without asteroids, where the orbital period is a commensurate fraction of that of Jupiter.

Lagrangian points   In any two-body gravitating system in which the primary and secondary are substantially different in mass, there are five gravitationally stable points where tiny masses (relative to the primary and secondary) can maintain stationary positions relative to the primary and secondary. L_1, L_2, and L_3 are found along the line connecting the primary and secondary, with L1 on the far side of the primary as seen by the secondary, L_2 between the primary and secondary, and L_3 on the far side of the secondary as seen by the primary. The L_4 and L_5 points are 60 degrees ahead of or behind, respectively, the secondary in its orbit around the primary.

L_3   The Lagrangian point along the line from primary mass through the secondary mass and behind the secondary as seen by the primary.

L_4   The Lagrangian point 60 degrees ahead of the secondary in its orbit around the primary.

L_5   The Lagrangian point 60 degrees behind the secondary in its orbit around the primary.

Meridian   The great circle in the sky passing from the north point on the horizon, through the observer's zenith, to the south point on the horizon.

Meteor   An event characterized by a rapidly moving point or streak of light caused by a particle as it disintegrates in the Earth's atmosphere.

Meteorite   A natural particle reaching the surface of the Earth from space after traveling through the atmosphere.

Meteoroid   A natural particle in space or in a planetary atmosphere.

Micrometeorite   A microscopic meteorite, often not strongly affected by its trip through the Earth's atmosphere.

Nongravitational forces   Forces changing a cometary orbit that are not due to gravitational effects; usually identified with jet forces on the nucleus (the so-called 'rocket effect').

Nucleus   The source of all cometary phenomena, the nucleus is believed to be a dirty snowball, or possibly a snowy dirtball, of frozen gases and dust.

Occultation   An event in which one body moves in front of another as seen by an observer. The covering body is substantially larger than the body being hidden.

Opposition   The point in the sky directly opposite the Sun. More generally, opposition refers to the position of an object on the opposite side of the sky from the Sun, though it may not be 180 degrees from the Sun. An object in opposition is generally as close as it can be to Earth during that apparition.

P/   A prefix added to a comet's name to indicate it is periodic.

PA   Position angle, measured from north (0 = 360 degrees) around through east (90 degrees), south (180 degrees), and west (270 degrees) and back to north.

Parent molecules   Water (H_2O), carbon dioxide (CO_2), hydrogen cyanide (HCN), and other molecules containing carbon and sulfur are believed to be the source molecules for many of the neutral and ionized atomic and molecular species observed in the coma and tail of a comet.

Perihelion   The nearest point in an object's orbit around the Sun.

Periodic   Refers to comets having orbital periods less than 200 years.

Perturb, perturbation   Changes in the orbital motion of an object caused by the gravity of masses other than the Sun (usually major planets) or by other forces.

Plasma   A 'gas' of positive and negative ions.

Plasma tail   A different name for an ion tail.

Poynting–Robertson effect   The effect wherein the absorption and emission of light causes the smallest dust particles, on average, to spiral in towards the sun.

Prograde   Orbital motion from west to east as seen on the sky and counterclockwise as seen looking down from the north pole.

Radiant   The direction in the sky from which a meteor enters the atmosphere and from which shower meteors appear to radiate. Meteor shower names are derived from the name of the constellation in which the radiant resides on the day of maximum.

Radiation pressure   Electromagnetic radiation (e.g. light, infrared, x-rays, radio, ultraviolet) has the property of being able to transfer momentum – push materials away from the source of the radiation.

Radius vector   The line from the center of the Sun projecting through an object and beyond it.

Resonance   The selective response of any periodic system to an external stimulus of the same frequency as the natural frequency of the system. Commonly used to discuss the orbital relationships of asteroids and major planets.

Retrograde   Orbital motion from east to west as seen on the sky and clockwise as seen looking down from the north pole.

Scatter, scattering   Small particles (one micrometer to 1/10 millimeter in size) have the property of not simply reflecting light and making shadows but actually spreading light that illuminates them in all directions. In some situations forward-scattered light, seen coming from the same direction as the source and appearing where a shadow would be expected, is actually brighter than back-scattered ('reflected') light.

Secular resonance   An asteroid is located in a secular resonance if the rate of precession of its orbit's longitude of nodes, or longitude of perihelion, is a small rational fraction of that of one of the major planets, notably Jupiter. Precession refers to a slow angular change in the orientation of the orbit. It is referred to the longitude of the nodes, which is the angle between the position of the vernal equinox and the position of the ascending node, or the longitude of perihelion, which is the angle between the position of the vernal equinox and the position of the body's perihelion.

Shower   Meteors radiating from the same radiant over a well-defined period of time and having approximately the same orbits belong to a meteor shower.

Solar wind   Ionized gases carrying magnetic fields are blown off the Sun at speeds ranging from 200 to 1000 km/s (120 to 600 miles/s).

Sporadic   A meteor not belonging to a meteor shower.

Storm   An extremely high-rate meteor shower.

Stream   A collection of meteoroids which have similar orbits. If the stream intersects Earth's orbit it is the source of a meteor shower.

Striae   Narrow, rectilinear structures sometimes seen in the dust tail of a comet. They are made of fragments of particles released at the same time from the nucleus that later disintegrate.

Sublimate   Change state directly from solid to gas without going through a liquid phase.

Synchrones   The loci of particles released from the comet nucleus simultaneously. They are sometimes seen in the dust tail as straight or moderately curved structures.

Syndynames   The loci of particles in a comet's dust tail that are subjected to equal force.

Tail   A general term used to describe the ejecta (ions and dust) streaming out from a comet nucleus and carried away from the Sun.

Trail   The short-term luminous glow left in the path of a meteor. Also called a wake.

Train   A persistent visual phenomenon along the path of a meteor which may appear either bright or dark.

Triangulation The trigonometric process of determining the height and distance of a target without physically touching the target. By measuring the difference in angular position of a target from two or more sites and knowing the physical distance between those sites it is possible to compute the position of the target from those sites.

Trojan asteroids Asteroids found orbiting the Sun near the leading L_4 or trailing L_5 Lagrangian points of a planet's orbit. Until the discovery of the Mars Trojan 1990 MB, Trojan asteroids referred only to the asteroids sharing Jupiter's orbit.

UT Universal Time, also called Greenwich Mean Time or Zulu. Time zones are defined relative to the Greenwich meridian which defines UT.

Wake The short-term luminous glow left in the path of a meteor. Also called a trail.

WWV, WWVH Stations broadcasting radio time signals for the US National Institute of Standards and Technology on frequencies of 2.5, 5, 10, 15, and 20 MHz (WWV) and 2.5, 5, 10, and 15 MHz (WWVH) 24 hours a day.

ZHR, Zenithal hourly rate The number of meteors per hour seen by an observer in perfect conditions with the meteor radiant at the zenith and a limiting magnitude of 6.5.

Zodiacal band This faint glow seen along the ecliptic connects the zodiacal light pyramids to the gegenschein.

Zodiacal light A general glow throughout the sky caused when sunlight is scattered by interplanetary dust. It is brightest near the Sun and along the ecliptic. The zodiacal light pyramids are often referred to simply as the zodiacal light.

Zodiacal light pyramid This triangular glow seen on the western horizon after evening twilight and the eastern horizon before morning twilight is the brightest component of the zodiacal light.

Appendix II: Report forms

COMET OBSERVING REPORT FORM. *Send completed sheets to INTERNATIONAL COMET QUARTERLY, c/o D. W. E. Green; Smithsonian Astrophysical Observatory; 60 Garden Street; Cambridge, MA 02138, U.S.A. Please convert observing times to decimals of a day in Universal Time (e.g., Aug. 3, 18:00 UT=Aug. 3.75 UT).*

Photocopy this sheet for more copies. (This form supercedes all previous forms. 1984 February 1.)

<u>PLEASE PRINT OR TYPE ONLY.</u> Use only one sheet per comet.

Name and designation of comet: COMET .... 19 ....

Observer .... Address ....

NOTE: Drawings and additional comments or remarks should be included on separate sheets of paper. To be eligible for publication in the ICQ, columns below marked with an asterisk (*) must be filled in.

Date* (U.T.)	M.* M.	Total* Magn.	Ref.*	Instr.* Aperture	Instr.* Type	f/*	Power*	Coma Dia.(')	D. C.	Tail Length (°)	P. A.	Remarks

Shower Observed _____

Visual/Radio Meteor Observation Report Form

UT Date: _____ Observer: _____

Dark Adaptation Time: _____ Site: _____

Count Method: Written: _____ Counter: _____ Tape Recorder: _____

Group Observation? No: _____ Yes: _____ **#** _____ List names of observers in Notes (below)

Viewing Area of Sky: Unrestricted _____ Limited to _____ ° x _____ °

UT Start – End	Limiting Mag.	Cloud Cover – %*	Facing Direction	Number of Meteors	
				Shower	Non-shower
			N E Z W S		
			N E Z W S		
			N E Z W S		
			N E Z W S		
			N E Z W S		
			N E Z W S		
			N E Z W S		
			N E Z W S		

*Indicate changes in percent of sky covered between Start and End times in the Notes section.

Notes (continue on reverse if necessary):

I.G.Y. Visual Meteor Report

Station

Date

Time

On [   ] Off   Pos.

Time

Time

No.	Mag	SH	Remarks
1			
2			
3			
4			
5			
6			
7			
8			
9			
10			
11			
12			
13			
14			
15			
16			
17			
18			
19			
20			
21			
22			
23			
24			
25			
26			
27			
28			
29			
30			

Time

No.	Mag	SH	Remarks
31			
32			
33			
34			
35			
36			
37			
38			
39			
40			
41			
42			
43			
44			
45			
46			
47			
48			
49			
50			
51			
52			
53			
54			
55			
56			
57			
58			
59			
60			
61			
62			
63			
64			
65			

Time

No.	Mag	SH	Remarks
66			
67			
68			
69			
70			
71			
72			
73			
74			
75			
76			
77			
78			
79			
80			
81			
82			
83			
84			
85			
86			
87			
88			
89			
90			
91			
92			
93			
94			
95			
96			
97			
98			
99			
100			

Meteor Section
Royal Astronomical Society of New Zealand

Name: ————————— Date: ————————— Began: —————————U.T.

Address: ——————————

Ended: —————————U.T.

——————————— Duration: ————————— Hrs ————————— Mins

——————————— Breaks: ————————— Hrs ————————— Mins

Limiting Mag: ————————— At Beginning: ————————————————— At End

Cloud Cover: ————————— At Beginning: ————————————————— At End

Shower(s) being observed: —————————————————————————————————

Time U.T.	Mag	Duration Meteor	Train	Length	Shower Sporadic	ACC	Notes (any peculiarities, etc)

Total: Sporadic: Shower:

FIREBALL REPORT

YEAR MONTH DAY

HOUR MINUTE am pm TIME ZONE

OBSERVER

WEATHER

ADDRESS

LOCATION
OF
OBSERVER
WHEN
FIREBALL
SEEN

BURSTS

LUMINOSITY

COLOUR

LAT LONG

FORM

DURATION

SOUNDS

POSITION
IN SKY

BEGIN

END

ELEVATION BEARING

DATE

RELIABILITY

ACM - 1

PLACE

REPORTER

METEOR SECTION

Royal Astronomical Society of New Zealand

FIREBALL REPORT SHEET

Please fill out as completely as possible and return promptly. Use the back of the sheet if necessary, numbering information according to the question on the front.

1. Observer ....
2. Mailing Address .... Phone ....
   ....
3. Observer location (where meteor was sighted) Geographic co-ordinates would be appreciated ....
   ....
4. Date .... Time .... Type of Time ....
   Weather conditions ....
5. Total duration of meteor was .... seconds.
6. In what direction did the meteor first appear (or was seen by you)? Please give compass points if you are not familiar with azimuth, where: 0°=North, 90°=East, 180°=South, 270°=West. ....
7. In what direction did the meteor disappear (or was last seen)? ....
8. At what angular height above the horizon (or angle from the zenith if you prefer) did the meteor first appear? Please indicate either horizon or zenith reference. ZENITH/HORIZON angle ....
9. At what angular height above the horizon (or from the zenith) did the meteor disappear? ZENITH/HORIZON angle ....
** For questions (6) through (9), you can supplement your answer by including (a) a map of the meteor's path among the stars (b) a sketch of the path with respect to known and readily identifiable land marks at the observing location.
10. Did it pass directly overhead? ....
11. If not, which side of the zenith did it appear? ....
12. Did it appear to reach the horizon? ....
13. Describe and/or sketch the horizon in the vicinity of the end point.
    ....
14. What angle did the meteor path make with the horizon near its end point?
    ....
15. Was the beginning point near the horizon? ....
    If so, what angle did path at start make with the horizon? ....
    Describe horizon near starting point. ....
16. Sometimes an extremely bright meteor explodes or leaves a visible train or wake. Did this object have a visible train? ....
    How long did the train last? ....
    (If it lasted long enough to drift, please include a series of sketches at various times showing carefully the shape and drift of the train with time relative to either the stars, the horizon or established land marks.)

Observations of Telescopic Meteors

Observed During the Year ⎯⎯⎯⎯⎯⎯⎯⎯⎯⎯⎯⎯⎯⎯⎯⎯⎯ Sheet No. ⎯⎯⎯⎯ of ⎯⎯

Observer ⎯⎯⎯⎯⎯⎯⎯⎯⎯⎯⎯⎯⎯⎯⎯⎯⎯⎯ Obs. No. ⎯⎯⎯⎯

Station ⎯⎯⎯⎯⎯⎯⎯⎯⎯⎯⎯⎯⎯⎯⎯⎯⎯⎯⎯⎯⎯⎯

⎯⎯⎯⎯⎯⎯⎯⎯⎯⎯⎯⎯⎯⎯⎯⎯⎯⎯⎯⎯⎯⎯⎯⎯⎯

Telescopes Used ⎯⎯⎯⎯⎯⎯⎯⎯⎯⎯⎯⎯⎯⎯⎯⎯⎯⎯⎯

Time Used: U.T. or ⎯⎯⎯⎯⎯⎯⎯⎯⎯⎯⎯⎯⎯⎯ Standard Time

No.	DATE			Magn.	Telescope Position Epoch......	Dir. of Motion	Eyep. Field Dia.	Var. Star Field	Color	Vel.	Limit. Vis.	NOTES
	Month	Day	Hour									

Instructions for using the telescopic meteors form

Column 1: The meteor's serial number for the year: 1, 2, 3, etc.

Columns 2, 3, and 4: The month, day, hour and minute meteor appeared: e.g. "Apr. 22, – 22.51". Be sure to state time used in space at top of page. Reckon hours from 0 to 23, avoiding "AM" and "PM".

Column 5: The estimated stellar magnitude of the meteor, to the nearest HALF MAGNITUDE only.

Column 6: State hour and minute of right ascension and nearest whole degree of declination in six-figure notation: e.g. 2 hours 15 minutes R.A. and 40° north declination would be "021540". Underline declination if south. State epoch of coordinates, as 1900, 1950, etc.

Column 7: The direction of motion of the meteor across the field of view. Approximate figures to the nearest 90 (or 45) degrees are sufficient: e.g. 090 = east, 270 = west, 135 = southeast, all reckoned from north toward east. Refer to coordinate lines on variable star chart, if in use, to determine direction of motion of meteor.

Column 8: The diameter of the field of view of the eyepiece in use at the time of observation, in minutes (or degrees and minutes) of arc.

Column 9: The name of the variable star field being observed when the meteor was seen. Use I.A.U. three letter abbreviations, as "R Leo", "U Her", "RS UMa", etc.

Column 10: Estimated color of meteor. Use the following letters: W = white; R = red; Y = yellow. Explain any other notations in space at bottom of page.

Column 11: Velocity of the meteor: For velocity estimates use the following abbreviations: SW = swift; SL = slow; AV = average; VSW = very swift; VSL = very slow.

Column 12: The faintest star visible in your telescope at the time the observation was made.

Column 13: Anything of interest including train, curvature of path, etc. For train, if seen, use "TR".

Imaging information record form

1. Instrument focal length ———— f/ ———— Aperture ———— Type ————

2. Instrument focal length ———— f/ ———— Aperture ———— Type ————

3. Camera lens: focal length ———— f/ ————

4. Camera lens: focal length ———— f/ ————

CCD information ————————————————————————

Emulsion name ———————————————— ISO (ASA/DIN) ————

Hypersensitized in ———————————— at ———— °C °F for ———— hours.

Emulsion cooled to ———— °C °F.

Developed in ———————————— at ———— °C °F for ———— minutes.

Grating: ———— gr/mm   Blaze order ————   Aperture ————

Prism: apex angle ———— ° Glass type ————   Aperture ————

Spectroscopic method ————————————

Rotating shutter chop frequency & open/closed ratio ————————————

Exposures

Object name	Negative number	UT Date	UT Start	Instru- ment No.	Photo. Method	EFL	Filter	Duration	Site	Sky Qual

Appendix III: Working list of meteor streams

For those of you who need a more detailed list than the one we offer in Chapter 5, we present a 'Working List of Meteor Streams', reprinted by permission from Alan F. Cook (1973). This list contains a few showers that have been observed but once, and offers precise radiants and other more technical information not included in the earlier list.

In this list, shower characteristic times are based on the longitude of the Sun, rather than on date. This is a more precise method of defining the time when the Earth crosses the orbital plane of the meteor stream. While the date and hour of this crossing can vary because the year is not exactly 365 days long, the solar longitude is constant. An ephemeris giving the solar longitude (such as the *Astronomical Almanac*) can be used by precessing the solar longitude from date to the shower list epoch (1950.0) and then finding the date and time (fraction of a day) by interpolation when the Sun crosses the longitude of maximum activity. Be aware, though, that meteor showers are not necessarily bound by printed schedules. They are just not that predictable.

Evolutionary and Physical Properties of Meteoroids I.—Working List of Meteor Streams

| Name | Dates[a] | Max. | Longitude of Sun (1950) | | | | | Geocentric radiant | | | |
			Beginning (deg)	Half max. (deg)	Max. (deg)	Half max. (deg)	End (deg)	R.A. 1950 (deg)	Decl. 1950 (deg)	Velocity (kms⁻¹)	Sun (deg)
Quadrantids	Jan. 1–4	Jan. 3	280.8	282.5	282.7	282.9	283.4	230.1	+48.5	41.5	282.7
δ Cancrids	Jan. 13–21	Jan. 16	293		296		301	126	+20	28	296
Virginids	Feb. 3–Apr. 15		314				25	186	0	35	350
δ Leonids	Feb. 5–Mar. 19	Feb. 26	316		338		359	159	+19	23	338
Camelopardalids	Mar. 14–Apr. 7		353				17	118.7	+68.3	6.8	359.0
σ Leonids	Mar. 21–May 13	Apr. 17	1		27		52	195	− 5	20	28
δ Draconids	Mar. 28–Apr. 17		7				27	281	+68	26.7	14
χ Serpentids	Apr. 1–7		11				17	230	+18	45	14
μ Virginids	Apr. 1–May 12	Apr. 25	12		35		51	221	− 5	29	35
α Scorpiids	Apr. 11–May 12	May 3	21		42		51	240	−22	35	42
α Boötids	Apr. 14–May 12	Apr. 28	24		36		51	218	+19	20	36
φ Boötids	Apr. 16–May 12	May 1	26		40		51	240	+51	12	40
April Lyrids	Apr. 20–23	Apr. 22	30.7	31.2	31.7	32.2	32.7	271.4	+33.6	47.6	31.7
η Aquarids	Apr. 21–May 12	May 3	30	39	42.4	45	51	335.6	− 1.9	65.5	42.4
τ Herculids	May 19–June 14	June 3	58		72		83	228	+39	15	72
χ Scorpiids	May 27–June 20	June 5	65		74		89	247	−13	21	74
Daytime Arietids	May 29–June 19	June 7	67	71	76	83	88	44	+23	37	77
Daytime ζ Perseids	June 1–17	June 7	70	72	76	83	86	62	+23	27	78
Librids	June 8–9, 1937	June 8	77.6		78.2		78.4+	227.2	−28.3	16±2	78.2
Sagittariids	June 8–16, 1957–8	June 11	77		80		82	304	−35	52	80
θ Ophiuchids	June 8–16	June 13	77		82		85	267	−28	26.7	82
June Lyrids	June 11–21, 1969	June 16	79	81	84.5	87.5	90	278	+35	31±3	84.5
Daytime β Taurids	June 24–July 6	June 29	91	93	96	99	103	86	+19	30	96
Corvids	June 25–30, 1937	June 26	94.8	94.9	95.2	97.6	97.9	191.9	−19.1	10±2	95.9
June Boötids	June 28, 1916	June 28	97.5		97.6		97.7	219	+49	13.9	98
July Phoenicids	July 3–18	July 14	101		112		116	31.1	−47.9	47±3	109.6
o Draconids	July 7–24	July 16	104				121	271	+59	23.6	113
Northern δ Aquarids	July 14–Aug. 25	Aug. 12	111		139		152	339	− 5	42.3	139
Southern δ Aquarids	July 21–Aug. 29	July 29	118	121	125	129	155	333.1	−16.5	41.4	125

α Capricornids	July 15–Aug. 10	July 30	123	126			138	307	−10	22.8	127
Southern ι Aquarids	July 15–Aug. 25	Aug. 5	112	131			151	333.3	−14.7	33.8	131.0
Northern ι Aquarids	July 15–Sept. 20	Aug. 20	112	147			177	327	− 6	31.2	147
Perseids	July 23–Aug. 23	Aug. 12	120	139	138	141	150	46.2	+57.4	59.4	139.0
κ Cygnids	Aug. 9–Oct. 6	Aug. 18	136	145			193	286	+59	24.8	145
Southern Piscids	Aug. 31–Nov. 2	Sept. 20	158	177			219	6	0	26.3	177
Northern Piscids	Sept. 25–Oct. 19	Oct. 12	182	199			206	26	+14	29	199
Aurigids	Sept. 1, 1935	Sept. 1		157.9				84.6	+42.0	66.3	157.9
κ Aquarids	Sept. 11–28	Sept. 21	168	178			184	338	− 5	16.0	178
Southern Taurids	Sept. 15–Nov. 26	Nov. 3	172	220			244	50.5	+13.6	27.0	220.0
Northern Taurids	Sept. 19–Dec. 1	Nov. 13	176	230	206	240	249	58.3	+22.3	29.2	230.0
Daytime Sextantids	Sept. 24–Oct.5	Sept. 29	179	184			190	152	0	32.2	183.6
Annual Andromedids	Sept. 25–Nov. 12	Oct. 3	182	190		195	230	5 / 20	+ 8 / +34	23.2 / 18.2	190 / 228
Andromedids	Nov. 27, 1885	Nov. 27	246.6	246.65	246.7	246.75	246.8	25	+44	16.5	247
Orionids	Oct. 2–Nov. 7	Oct. 21	189	206.7	207.7	208.3	225	94.5	+15.8	66.4	208.0
October Draconids	Oct. 9	Oct. 9	196.25	196.3			196.35	262.1	+54.1	20.43	196.3
ε Geminids	Oct. 14–27	Oct. 19	201	206			214	104	+27	69.4	209
Leo Minorids	Oct. 22–24	Oct. 24	209	211			211	162	+37	61.8	211
Pegasids	Oct. 29–Nov. 12	Nov. 12	215	230			230	335	+21	11.2	230
Leonids	Nov. 14–20	Nov. 17	231	234.447	234.462	234.477	237	152.3	+22.2	70.7	234.5
Monocerotids	Nov. 27–Dec. 17	Dec. 10	245	258			265	99.8	+14.0	42.4	257.6
σ Hydrids	Dec. 3–15	Dec. 11	251	259			263	126.6	+ 1.6	58.4	259.0
Northern χ Orionids	Dec. 4–15	Dec. 10	252	258			261	84	+26	25.2	258
Southern χ Orionids	Dec. 7–14	Dec. 11	255	259			262	85	+16	25.5	259
Geminids	Dec. 4–16	Dec. 14	252	260.6	261.7	262.1	264.2	112.3	+32.5	34.4	261.0
December Phoenicids	Dec. 5, 1956	Dec. 5	253.18	253.45	253.55	253.65	253.70	15	−55 / −45	21.7 / 11.7	253
δ Arietids	Dec. 8–14		256				262	52	+22	13.2	254
Coma Berenicids	Dec. 12–Jan. 23		260				303	175	+25	65	257.6
Ursids	Dec. 17–24	Dec. 22	265	270		271	272	217.06	+75.85	33.4	270.66

Note: ᵃ Unless otherwise indicated, all calendar dates are for the year 1950.

Appendix IV: A simple reduction program for astrometry

By Jordan D. Marché II; adapted by Roger Sinnott for use with Quickbasic. Reprinted by permission of *Sky and Telescope* Magazine. The program is explained in that magazine's July 1990 issue, page 71.

When you copy this program into your computer, remember that these lines must not wrap around. If they do, the program won't run!

```
100 REM ASTROMETRIC REDUCTION
105 REM
106 DEFDBL A-H, K-Z
110 DIM X(10), Y(10), X1(10), Y1(10)
115 DIM a(10), D(10), R(10, 9), RA(10), RD(10)
125 PI = 3.141592653589793#: DR = PI / 180
130 INPUT "Camera focal length"; L
135 PRINT "R.A. of plate center (h,m,s)";
140 GOSUB 630: A0 = V * 15 * DR
145 PRINT "Dec. of plate center (d,m,s)";
150 GOSUB 630: D0 = V * DR: SD = SIN(D0): CD = COS(D0)
160 INPUT "Equinox, epoch"; EQ, EP
165 T = EP - EQ
170 INPUT "How many stars (4-10)"; N
175 FOR J = 1 TO N
180 PRINT
185 PRINT "R.A. of star"; J; "(h,m,s)";
190 GOSUB 630
195 INPUT "Proper motion (sec/yr)"; M1
200 a(J) = (V + T * M1 / 3600) * 15 * DR
205 PRINT "Dec. (d,m,s)";
210 GOSUB 630
215 INPUT "Proper motion (arcsec/yr)"; M2
220 D(J) = (V + T * M2 / 3600) * DR: SJ = SIN(D(J)): CJ = COS(D(J))
225 H = SJ * SD + CJ * CD * COS(a(J) - A0)
230 X1(J) = CJ * SIN(a(J) - A0) / H
235 Y1(J) = (SJ * CD - CJ * SD * COS(a(J) - A0)) / H
240 INPUT "Measured X,Y"; X(J), Y(J)
```

245 NEXT J

250 PRINT

255 INPUT "Measured X,Y of target"; X, Y

260 R1 = 0: R2 = 0: R3 = 0: R7 = 0: R8 = 0: R9 = 0: XS = 0: YS = 0

265 FOR J = 1 TO N

270 XS = XS + X(J): YS = YS + Y(J): R(J, 1) = X(J) * X(J)

275 R1 = R1 + R(J, 1): R(J, 2) = Y(J) * Y(J): R2 = R2 + R(J, 2)

280 R(J, 3) = X(J) * Y(J): R3 = R3 + R(J, 3)

285 R(J, 7) = Y1(J) − Y(J) / L: R7 = R7 + R(J, 7)

290 R(J, 8) = R(J, 7) * X(J): R8 = R8 + R(J, 8)

295 R(J, 9) = R(J, 7) * Y(J): R9 = R9 + R(J, 9)

300 NEXT J

305 REM Now solve for D, E, F, by Cramer's Rule

310 DD = R1 * (R2 * N − YS * YS) − R3 * (R3 * N − XS * YS) + XS * (R3 * YS − XS * R2)

315 D = R8 * (R2 * N − YS * YS) − R3 * (R9 * N − R7 * YS) + XS * (R9 * YS − R7 * R2)

320 E = R1 * (R9 * N − R7 * YS) − R8 * (R3 * N − XS * YS) + XS * (R3 * R7 − XS * R9)

325 F = R1 * (R2 * R7 − YS * R9) − R3 * (R3 * R7 − XS * R9) + R8 * (R3 * YS − XS * R2)

330 D = D / DD: E = E / DD: F = F / DD

335 REM

340 R4 = 0: R5 = 0: R6 = 0

345 FOR J = 1 TO N

350 R(J, 4) = X1(J) − X(J) / L: R4 = R4 + R(J, 4)

355 R(J, 5) = R(J, 4) * X(J): R5 = R5 + R(J, 5)

360 R(J, 6) = R(J, 4) * Y(J): R6 = R6 + R(J, 6)

365 NEXT J

370 REM Now solve for A ,B, C, by Cramer's Rule

375 a = R5 * (R2 * N − YS * YS) − R3 * (R6 * N − R4 * YS) + XS * (R6 * YS − R4 * R2)

380 B = R1 * (R6 * N − R4 * YS) − R5 * (R3 * N − XS * YS) + XS * (R3 * R4 − XS * R6)

385 C = R1 * (R2 * R4 − YS * R6) − R3 * (R3 * R4 − XS * R6) + R5 * (R3 * YS − XS * R2)

390 a = a / DD: B = B / DD: C = C / DD

395 PRINT

400 PRINT "   Plate Constants"

405 PRINT "   R.A.   Dec."

410 PRINT USING "A = ##.#####   D = ##.#####"; a; D

415 PRINT USING "B = ##.#####   E = ##.#####"; B; E

420 PRINT USING "C = ##.#####   F = ##.#####"; C; F

```
425 REM
430 REM   NOW FIND RESIDUALS
435 BS = 0: DS = 0
440 FOR J = 1 TO N
445 RA(J) = X(J) − L * (X1(J) − (a * X(J) + B * Y(J) + C))
450 RD(J) = Y(J) − L * (Y1(J) − (D * X(J) + E * Y(J) + F))
455 BS = BS + ((RA(J) / L) * 3600 / (DR * 15 * COS(D0))) ∧ 2
460 DS = DS + ((RD(J) / L) * 3600 / DR) ∧ 2
465 NEXT J
470 S1 = SQR(BS / (N − 3)): S2 = SQR(DS / (N − 3))
475 PRINT
480 PRINT "Residuals              R.A.          Dec."
482 a$ = "Star ##          #.#####      #.#####"
485 FOR J = 1 TO N
490 PRINT USING a$; J; RA(J); RD(J)
495 NEXT J
500 PRINT
505 REM Find standard coordinates of target
510 XX = a * X + B * Y + C + X / L: YY = D * X + E * Y + F + Y / L
515 B = CD − YY * SD: G = SQR(XX * XX + B * B)
520 REM
525 REM Find right ascension of target
530 A5 = ATN(XX / B): IF B < 0 THEN A5 = A5 + PI
535 A6 = A5 + A0: IF A6 > 2 * PI THEN A6 = A6 − 2 * PI
540 IF A6 < 0 THEN A6 = A6 + 2 * PI
545 V = A6 / (DR * 15): GOSUB 660: A1 = V1: A2 = V2: A3 = V3
550 REM
555 REM Find declination of target
560 D6 = ATN((SD + YY * CD) / G): V = D6 / DR: GOSUB 660
570 D1 = V1: D2 = V2: D3 = V3
575 REM
580 PRINT "For target:"
585 PRINT
590 PRINT "Right ascension      ";
592 PRINT USING "## ## ##.###"; A1; A2; A3
595 PRINT USING "    Std. dev.                    ##.###"; S1
600 PRINT
605 PRINT "Declination          "; S$;
610 PRINT USING "## ## ##.##"; D1; D2; D3
615 PRINT USING "    Std. dev.                    ##.##"; S2
620 END
625 REM
630 REM Input of sexagesimal values
```

```
635 INPUT V$, V2, V3
640 S = 1: IF LEFT$(V$, 1) = "−" THEN S = −1
645 V1 = ABS(VAL(V$)): V = S * (V1 + V2 / 60 + V3 / 3600)
655 RETURN
660 REM Output of sexagesimal values
665 S$ = "+": IF V < 0 THEN S$ = "−"
670 V = ABS(V): V1 = INT(V): VM = 60 * (V − V1)
675 V2 = INT(VM): V3 = 60 * (VM − V2)
680 RETURN
690 REM
700 REM This program is used to analyze measurements
710 REM of minor planet or comet positions on a
720 REM photographic plate and deduce precise
730 REM coordinates. Written by Jordan D. Marche
740 REM and explained by him in Sky & Telescope
750 REM for July, 1990, page 71.
```

Appendix V: Addresses of organizations and publications

AAS, American Astronomical Society, Executive Office, 2000 Florida Avenue, NW, Suite 300, Washington, DC 20009.

Acta Astronomica, Copernicus Foundation for Polish Astronomy, ul Ujazdowskie 4, PL 00-478 Warsaw, Poland.

American Association of Variable Star Observers (AAVSO), 25 Birch Street, Cambridge, MA 02138, USA.

American Meteor Society, c/o Prof. David Meisel, Dept. of Physics and Astronomy, State University of New York at Geneseo, Geneseo, NY 14454, USA.

Association of Lunar and Planetary Observers (ALPO), *Journal of the ALPO (The Strolling Astronomer)*, P.O. Box 16131, San Francisco, CA 94116, USA.
   Comets Section: Don Machholz, P.O. Box 1716, Colfax, CA 95713, USA.
   Meteors Section: Robert D. Lunsford, 161 Vance Street, Chula Vista, CA 91910, USA.

Astronomical Almanac, US Naval Observatory, 34th and Massachusetts Ave. N.W., Washington, DC 20390, USA.

Astronomical Journal, Editorial Office, Dept. of Astronomy, FM-20, University of Washington, Seattle, WA 98195, USA.

Astronomical League, Merry Edenton-Wooten, Executive Secretary, 6235 Omie Circle, Pensacola, FL 32504, USA.

Astronomical Society of the Pacific (ASP), *Mercury*, 390 Ashton Avenue, San Francisco, CA 94112, USA.

Astronomy, Kalmbach Publishing, 21027 Crossroads Circle, P.O. Box 1612, Waukesha, WI 53187, USA.

Astrophysical Journal, Editorial Office, Kitt Peak National Observatory, PO Box 26732, Tucson, AZ 85726-6732, USA.

Australian Comet Section, David Seargent, 156 Entrance Rd., The Entrance, NSW 2261, Australia.

BAA Comet Section, Jonathan Shanklin, 11 City Road, Cambridge, CB1 1DP, UK.

BAA Handbook, British Astronomical Association, Burlington House, Piccadilly, London, W1V 0NL, UK.

224

Central Bureau for Astronomical Telegrams, Smithsonian Astrophysical Observatory, 60 Garden Street, Cambridge, MA 02138, USA. Telephone 617–495–7244/7440/7444 (for emergency use only) TWX 710–320–6842 ASTROGRAM CAM, EASYLINK 62794505, e-mail MARSDEN@CFA or GREEN@CFA (followed by the suffix .SPAN, .BITNET or .HARVARD.EDU)

Comet and Minor Planet Section, RASNZ, Alan C. Gilmore, Mt. John University Observatory, P.O. Box 20, Lake Tekapo, South Canterbury, NZ.

The *Comet Handbook* is published annually by the *International Comet Quarterly*, listed below.

Comet Rapid Announcement Service including *CRAS Notices* and the *Shallow Sky Bulletin*, P.O. Box 110282, Cleveland, OH 44111–0282, USA.

(Dutch Comet Section) NVWS Werkgroep Kometen, E. Bakker, Brabantse Turfmarkt 83a, 2611 CM Delft, The Netherlands.

Hungarian Comet Observing Network, K. Sárneczky, Kádár u. 9–11. fsz.3., 1132 Budapest, Hungary.

Hungarian Meteor Section of the Hungarian Astronomical Association, I. Tepliczky, Baji ut 42, H-2890 Tata, Hungary.

Icarus, Editorial Office, Space Sciences Building, Cornell University, Ithaca, NY 14853-6801, USA.

International Amateur–Professional Photoelectric Photometry Association (IAPPP), Prof. Douglas Hall, Dyer Observatory, Vanderbilt University, Nashville, TN 37235, USA.

International Comet Quarterly, Daniel Green, Smithsonian Astrophysical Observatory, 60 Garden St., Cambridge, MA 02138, USA.

International Meteor Organization, Paul Roggemans, Pijnboomstraat 25, B-2800 Mechelen, Belgium.
  Fireball Data Center, c/o Andre Knoefel, Saarbrueker Str. 8, D-W–4000 Düsseldorf 30, Germany.

International Occultation Timing Association and *Occultation Newsletter*, c/o Craig and Terry McManus, 2760 SW Jewell Avenue, Topeka, KS 66611–1614, USA.

(Japanese Amateur Comet Observers), Hoshino Hiroba, Akira Kamo, 5–10 Shimazaki-icho, Wakayama-shi 640, Japan.

Meteor News, Wanda Simmons, Callahan Astronomical Society, Rt. 3, Box 1062, Callahan, FL 32011, USA.

Microsky, Dccn Publications, P.O. Box 867088, Plano, TX 75086–7088, USA.

Minor Planet Bulletin; contributions to Richard Binzel, Dept. Earth, Atmos. Plan Sci., 77 Massachusetts Ave., Cambridge, MA 02139, USA and subscription information requests to Derald Nye, 10385 E. Observatory Drive, Tucson, AZ 85747, USA.

225

Minor Planet Circulars (or *Minor Planets and Comets*), Smithsonian Astrophysical Observatory, 60 Garden St., Cambridge, MA 02138, USA.

Nature, Macmillan Magazines Ltd, 4 Little Essex Street, London WC2R 3LF, UK.

NSSDC, Goddard Spaceflight Center, Greenbelt, MD 20771, USA.

Palomar Observatory, Office of the Director, Caltech 105–24, Pasadena, CA 91125, USA.

Physics Today, 335 East 45 Street, New York, NY 10017, USA.

Popular Science, 2 Park Avenue, New York, NY 10016, USA.

QST, American Radio Relay League, 225 Main Street, Newington, CT 06111-1494, USA.

RASC *Observer's Handbook,* Royal Astronomical Society of Canada, 124 Merton St., Toronto, Ontario, M4S 2Z2, Canada.

Royal Astronomical Society, Burlington House, Piccadilly, London, V1V ONL, UK.

Royal Astronomical Society of Canada, 136 Dupont Street, Toronto, Ontario M5R 1V2, Canada.

Scientific American, Inc., 415 Madison Avenue, New York, NY 10017-1111, USA.

Sky and Telescope and Sky Publications, Sky Publishing Corp., P.O. Box 9111, Belmont, MA 02178–9111, USA.

The *Sky-Gazer's Almanac* is in the January *Sky and Telescope* each year. The address of *Sky and Telescope* is given above.

Smithsonian Astrophysical Observatory, Harvard-Smithsonian Center for Astrophysics, 60 Garden Street, Cambridge, MA 02138, USA.

Smithsonian Institution, Smithsonian Associates, 900 Jefferson Drive, Washington, DC 20560, USA.

Space Telescope Science Institute, Hopkins Homewood Campus, Baltimore, MD 21218, USA.

(Spanish Society of Meteor and Comet Observers), Sociedad de Observadores Españoles de Meteores y Cometas, Avinguda de l'Antic Regne de València 35–9ª pta, 46005 València, Spain.

Appendix VI: References and bibliography

Adams, M. T., 'The Influence of Sky Conditions on Observed Meteor Rates', preprint.

Arbour, R. W., 1985, 'An Amateur's Computerised Camera for the Automatic Tracking of Comets', *Journal of the BAA*, Vol. 96, No. 1, p. 12.

'Asteroid Streams' in 'News Notes', *Sky and Telescope*, 1991 May, p. 467.

Astronomical Phenomena for 1993, Washington: US Naval Observatory.

Bain, W. F., 1957 April, 'V.H.F. Meteor Scatter Propagation', *QST*, pp. 20, 140, 142, 144.

Bain, W. F., 1974 May, 'VHF Propagation by Meteor-Trail Ionization', *QST*, pp. 41, 176.

Balbi, L., 1987 September, 'A Wire Micrometer for Photographs' in 'Gleanings for ATM's', *Sky and Telescope*, p. 310.

Beish, J. D. and Capen, C. F., 1988, *Mars Observer's Handbook*, Pasadena: The Planetary Society.

Berry, R., 1991, *Introduction to Astronomical Image Processing*, Richmond: Willmann-Bell.

Berry, R., 1992, *Choosing and Using a CCD Camera*, Richmond: Willmann-Bell.

Black, W. H., 1983 July, 'Observing Meteors By Radio' in 'Amateur Astronomers', *Sky and Telescope*, p. 61.

Blackwell, S. K., 1992 October, 'Make Your Own Dew Eliminator' in 'Gleanings for ATM's', *Sky and Telescope*, p. 455.

Bobrovnikoff, N. T., 1931, 'Halley's Comet in Its Apparition of 1909–1911', *Publications of the Lick Observatory*, Univ. of California Press, Vol. 17, Pt. II, p. 305.

Bortle, J. E., 1980, 'How to Observe Comets' in the *'Sky and Telescope' Guide to the Heavens*, Robinson, L. J., ed. (also published in *Sky and Telescope*, March 1981, p. 210), Cambridge, MA: Sky Publishing Corp.

Bradfield, W. A., 1981 July, 'Source Procedures for Comet Discovery', *International Comet Quarterly*, p. 71.

Brandt, J. C., 1981, *Comets – Reading from Scientific American*, San Francisco: W. H. Freeman & Co.

Brandt, J. C. and Chapman, R. D., 1981, *Introduction to Comets*, Cambridge: Cambridge Univ. Press.

Brandt, J. C., Donn, B., Greenberg, J. M., and Rahe, J., eds., 1981, *Modern Observational Techniques for Comets*, Pasadena: NASA-JPL Pub. 81–68.

Brandt, J. C., Friedman, L. D., Newburn, R. L., and Yeomans, D. K., eds., 1980, *The International Halley Watch – Report of the Science Working Group*, Pasadena: NASA – JPL Publication TM 82181 (400–88).

Brown, P. L., 1973, *Comets, Meteorites and Men*, New York: Taplinger Pub. Co.

Calder, N., 1981, *The Comet is Coming*, New York: Viking Press.

Canady, W., 1991 October, 'A Novel Nighttime Star Pointer' in 'Gleanings for ATMs', *Sky and Telescope*, p. 415.

Chapman, C. R., 1993 June 10, 'Comet on target for Jupiter', *Nature*, Vol. 363, p. 14.

Chou, B. R., 1992 November, 'How Filtered Glasses Help Dark Adaptation' in 'Observer's Page', *Sky and Telescope*, p. 582.

'Cloud Satellite of the Earth' in 'News Notes', *Sky and Telescope*, 1961 December, p. 328.

Cook, A. F., 1973, 'Working List of Meteor Streams', in *Evolutionary and Physical Properties of Meteoroids*, Hemenway, C. L., Millman, P. M., and Cook, A. F., eds., Washington, DC: NASA, SP-319, p. 183.

Croswell, K., 1991 November, 'Will the Lion Roar Again?', *Astronomy*, p. 44.

Cunningham, C. J., 1988, *Introduction to Asteroids*, Richmond, VA: Willmann–Bell, Inc.

Dahlmark, L., 1988, 'Searching for Variables, and Constructing a Slit-less Spectroscope' (abstract only) in *Stargazers – The Contributions of Amateurs to Astronomy (Proc. IAU Colloquium 98)*, Dunlop, S. and Gerbaldi, M., eds., Berlin: Springer-Verlag, p. 111.

'Designations of Magnitude References', *International Comet Quarterly*, 1981 April, p. 47.

de Vaucouleurs, G., 1961, *Astronomical Photography*, London: Faber and Faber.

Drummond, J. D., 1991, 'Earth Approaching Asteroid Streams', *Icarus*, Vol. 89, No. 1, p. 14.

Dubiago, A. D., 1961, *The Determination of Orbits*, New York: MacMillan (translated by R. D. Burke *et al*).

Dunham, D. W., Dunham, J. B., Binzel, R. P., Evans, D. S., Freuh, M., Henry, G. W., A'Hearn, M. F., Schnurr, R. G., Betts, R., Haynes, H., Orcutt, R., Bowell, E., Wasserman, L. H., Nye, R. A., Giclas, H. L., Chapman, C. L., Dietz, R. D., Moncivais, C., Douglass, W. T., Parker, D. C., Beish, J. D., Martin, J. O., Monger, D. R., Hubbard, W. B., Reitsema, H. J., Klemola, A. R., Lee, P. D., McNamara, B. R., Maley, P. D., Manly, P., Markworth, N. L., Nolthenius, R., Oswalt, T. D., Smith, J. A., Strother, E. F., Povenmire, H. R., Purrington, R. D., Trenary, C., Schneider, G. H., Schuster, W. J., Moreno, M. A., Guichard, J., Sanchez, G. R., Taylor, G. E., Upgren, A. R., and Van Flandern, T. C., 1990, 'The Size and Shape of (2) Pallas From the 1983 Occultation of 1 Vulpeculae', *Astronomical Journal*, Vol. 99, No. 5, p. 1636.

'Earth's Cloud Satellites Reported Again' in 'News Notes', *Sky and Telescope*, 1991 February, p. 132.

Edberg, S. J., 1982 March, 'Exploring the Stars with a Spectrograph' in 'Observer's Page', *Sky and Telescope*, p. 311.

Edberg, S. J., 1983, *International Halley Watch Amateur Observers' Manual for Scientific Comet Studies*, Pasadena: NASA-JPL Publication 83–16.

Eicher, D. J., 1983 February, 'Capturing Deep Sky Objects On Paper' in 'Gazer's Gazette', *Astronomy*, p. 35.

Everhart, E., 1982 September, 'Constructing a Measuring Engine' in 'Gleanings for ATM's', *Sky and Telescope*, p. 279.

Feibelman, W. A., 1974 August, 'A Well-Observed Bright Perseid Meteor' in 'Observer's Page', *Sky and Telescope*, p. 127.

Feijth, H., 1980 October, 'Of Sequences and Comparison Star Magnitudes', *International Comet Quarterly*, p. 73.

Fox, J., 1982, *Math For Amateur Astronomers*, Peoria: Astronomical League.

Genet, R. M., 1983a February, 'Backyard Photoelectric Photometry', in 'Equipment Atlas', *Astronomy*, p. 51.

Genet, R. M., ed., 1983b, *Solar System Photometry Handbook*, Richmond: Willmann-Bell.

González, D. D., 1993 Enero – Febrero, 'Fundamentos Para el Análisis Digital de Reflexiones Meteóricas', *Meteors – Circular Informativa*, No. 25, p. 14.

Greene, C., 1986 January, 'Meteor-Scatter Communications', *QST*, p. 14.

Guillemin, A., 1877, *The World of Comets*, London: Sampson Low.

Hall, D. S. and Genet, R. M., 1981, *Photoelectric Photometry of Variable Stars*, Fairborn, OH: Fairborn Observatory.

'Halley's Spin Period is 10hr 19min', *Comet News Service*, McDonnell Planetarium, St. Louis, No. 80–2, 1980 April 24.

Halliday, I., Blackwell, A. T., Griffin, A. A., 1990, 'Evidence for the Existence of Groups of Meteorite-Producing Asteroidal Fragments', *Meteoritics*, Vol. 25, p. 93.

Hamid, S. E., and Whipple, F. L., 1963, 'Common Origin Between the Quadrantids and the Delta Aquarids Streams', *Astronomical Journal*, Vol. 68, No. 8, p. 537.

Hardorp, J., 1982, 'The Sun Among the Stars V. A Second Search for Solar Spectral Analogs. The Hyades Distance', *Astron. Astrophys.* Vol. 105, p. 120.

Harris, S., 1967 January, 'November Leonids – Shower of a Lifetime', in 'The World Above 50 Mc.', *QST*, p. 83.

Harvey, G. A., 1974 June, 'Four Years of Meteor Spectra Patrol', *Sky and Telescope*, p. 378.

Henden, A. and Kaitchuck, R., 1982, *Astronomical Photometry*, New York: Van Nostrand Reinhold. (A new printing is available from Willmann–Bell in Richmond, VA.)

Hendrie, M., 1984 September, 'IHW Astrometry Network Workshop – 1984 June 18–19', *British Astronomical Association Comet Section Bulletin*, No. 22, p. 11.

Herald, D., 1981 April, 'Visual Magnitudes and the SAO Catalog', *International Comet Quarterly*, p. 43.

Hindley, K. B., 1972 March, 'The Quadrantid Meteor Stream', *Sky and Telescope*, p. 162.

Hodgson, R. G., 1962 May–June, 'Report on Dr. K. Kordylewski's 'Cloud Satellites:' A Negative Observation of the L4 Position', *The Strolling Astronomer, Journal of the Association of Lunar and Planetary Observers*, p. 99.

Hoffleit, D. and Jaschek, C., 1982, *The Bright Star Catalog*, New Haven: Yale Univ.

Houston, W. S., 1958 July, 'How a Kansas Amateur Group Counts Meteors By Reflection Of Radio Waves' in 'The Amateur Scientist', conducted by C. L. Stong, *Scientific American*, p. 108.

Hughes, D. W., 1978, 'Meteors', in *Cosmic Dust*, McDonnell, J. A. M., ed., New York: John Wiley & Sons.

Huling, John Jr., 1981 July, letter to the editor in 'Letters', *Sky and Telescope*, p. 16.

Jones, Rev. G. J., 1856, *Observations on the Zodiacal Light, From April 2, 1853, to April 22, 1855, Made Chiefly on Board the United States Steam-Frigate Mississippi, During Her Late Cruise in Eastern Seas, and Her Voyage Homeward: With Conclusions From the Data Thus Contained*, Washington: Senate Printer.

Keay, C. S. L., 1985 December, 'In Search of Meteor Sounds' in 'Observer's Page', *Sky and Telescope*, p. 623.

Keay, C., 'Meteor Fireball Sounds Identified' in *Abstracts for The International Conference on Asteroids, Comets, Meteors 1991*, Houston: Lunar and Planetary Institute, LPI Contribution No. 765, p. 111. (See also the proceedings of the conference published the following year.)

Kimball, D. S., 1957 May, 'Visual Aurora Observing During the IGY', *Sky and Telescope*, p. 327.

King, E. S., 1931, *A Manual of Celestial Photography*, Boston: Eastern Science Supply Co. (A new printing is available from Sky Publishing Corp. in Cambridge, MA)

Kodak Filters for Scientific and Technical Uses, (updated at irregular intervals), Rochester: Kodak Publication B-3.

Kordylewski, K., 1961, 'Photographische Untersuchungen des Librationspunktes L_5 im System Erde–Mond', *Acta Astronomica*, Vol. 11, p. 165.

Kronk, G. W., 1984, *Comets: A Descriptive Catalog*, Hillside, NJ: Enslow.

Kronk, G. W., 1988, *Meteor Showers: A Descriptive Catalog*, Hillside, NJ: Enslow.

Lacroix, D. P., 1982 July, 'Stellar Spectra the Easy Way' in 'Observer's Page', *Sky and Telescope*, p. 99.

Lampkin, R. H., 1972, *Naked Eye Stars*, Edinburgh: Gall and Inglis.

Landolt, A. U., 1973, 'UBV Photoelectric Sequences in the Celestial Equatorial Selected Areas 92 – 115', *Astronomical Journal*, Vol. 78, p. 959.

Larson, S. M., Edberg, S. J., and Levy, D. H., 1991, 'The Modern Role of Visual Observations of Comets' in Newburn, R. L., Jr., Neugebauer, M., and Rahe, J., *Comets in the Post-Halley Era*, Vol. I, Dordrecht: Kluwer Academic Publishers, p. 209.

Lasker, B. M., Sturch, C. R., Lopez, C., Mallamas, A. D., McLaughlin, S. F., Russell, J. L., Wiśniewski, W. Z., Gillespie, B. A., Jenkner, H., Siciliano, E. D., Kenny, D., Baumert, J. H., Goldberg, A. M., Henry, G. W., Kemper, E., and Siegel, M. J., 1988 September, 'The Guide Star Photometric Catalog', *The Astrophysical Journal Supplement Series*, Vol. 68, No. 1.

Lazerson, H., 1984 March, 'A Binocular Blink Comparator' in 'Gleanings for ATM's', *Sky and Telescope*, p. 275.

Levison, H. F., Shoemaker, E. M., and Wolfe, R. F., 1991, 'Mapping the Stability Field of Trojan Orbits in the Outer Solar System' in *Abstracts for The International Conference on Asteroids, Comets, Meteors 1991*, Houston: Lunar and Planetary Institute, LPI Contribution No. 765, p. 135. (See also the proceedings of the conference published the following year.)

Levy, D. H., 1993 July, 'Pearls on a String', *Sky & Telescope*, p. 38.

Liller, W., 1986 May, 'Tracking a Comet' in 'Gleanings for ATM's', *Sky and Telescope*, p. 514.

Lines, R. D., 1973a June, 'A Simple Micrometer Microscope for Off-Axis Guiding in Comet Photography', *The Strolling Astronomer, Journal of the Association of Lunar and Planetary Observers*, p.97.

Lines, R. D., 1973b June, 'How to Use a Micrometer Microscope to Guide for Comet Photography', *The Strolling Astronomer, Journal of the Association of Lunar and Planetary Observers*, p. 98.

Littman, M. and Yeomans, D. K., 1985, *Comet Halley: Once in a Lifetime*, Washington, DC: American Chemical Society.

Lovell, A. C. B., 1954, *Meteor Astronomy*, London: Oxford Univ. Press.

Lucas, G. A., 1987, 'Dewing – What It Is and What to Do About It', in *RTMC 87*, Costa Mesa, CA: Orange County Astronomers.

Lynch, J. L., 1992 August, 'A Different Way to Observe the Perseids' in 'Observer's Page', *Sky and Telescope*, p. 222.

MacRobert, A. M., 1988 August, 'Meteor Observing – I' in 'Backyard Astronomy', *Sky and Telescope*, p. 131.

MacRobert, A. M., 1988 October, 'Meteor Observing – II' in 'Backyard Astronomy', *Sky and Telescope*, p. 363.

MacRobert, A. M., 1992 August, 'A Crucial Year for the Perseid Meteors' in 'Celestial Calendar', *Sky and Telescope*, p. 185.

Machholz, Don, 1985, *A Decade of Comets: A Study of the 33 Comets Discovered By Amateur Astronomers Between 1975 and 1984*. Private publ., D. Machholz.

Majden, E. P., 1978 June, 'Conventional Meteor Spectroscopy for Amateurs', *Meteor News*, No. 41, p. 1.

Marché, J. D., II, 1990 July, 'Measuring Positions on a Photograph', in 'Astronomical Computing', *Sky and Telescope*, p. 71.

Marcus, J. N., 1981 January 12, 'Observing Comets for Nuclear Rotation', *Comet News Service*, McDonnell Planetarium, St. Louis, No. 81–1.

Marsden, B.G., 1967, 'The Sungrazing Comet Group', *Astronomical Journal*, Vol. 72, p. 9.

Marsden, B. G., 1979, 'Comet Halley and History', *Space Missions to Comets*, Washington, DC: NASA, CP-2089.

Marsden, B. G., 1982 September, 'How to Reduce Plate Measurements' in 'Gleanings for ATMs', *Sky and Telescope*, p. 284.

Marsden, B. G., 1983, *Catalog of Cometary Orbits*, Hillside, NJ: Enslow. (An updated edition is available from the Central Bureau for Astronomical Telegrams.)

Marsden, B. G. and Roemer, E., 1982, 'Basic Information and References' in *Comets*, Wilkening, L. L., ed., Tucson: University of Arizona Press.

Mayall, R. N. and Mayall, M. W., 1968, *Skyshooting – Photography for Amateur Astronomers*, New York: Dover Publications, Inc.

Mayer, B., 1974 July, 'A Self-Operating Meteor Camera' in 'Gleanings for ATMs', *Sky and Telescope*, p. 54.

Mayer, B., 1977 September, 'Projection Blinking – A Way Toward Discovery' in 'Observer's Page', *Sky and Telescope*, p. 246.

Mayer, B., 1978 May, 'Blink for a Nova' in 'Photography in Astronomy', *Astronomy*, p. 34.

Mayer, B., 1984, *Starwatch*, New York: Putnam (Perigee) Publishing Group.

Mayer, B., 1988, *Astrowatch*, New York: Putnam Publishing Group.

McKinley, D. W. R., 1961, *Meteor Science and Engineering*, New York: McGraw-Hill Book Co., Inc.

McLeod, N. W., III, *Visual Meteor Observing – Notes and Instructions*, American Meteor Society, State Univ. of New York, Geneseo.

Meisel, D. D., 'Radiowave Scatter Detection of Meteors Using VHF Aeronautical Beacons', an American Meteor Society bulletin.

'Meteor Monitoring Is Meaningful' in 'Amateur Astronomers', *Sky and Telescope*, 1978 January, p. 37.

Middlehurst, B. M. and Kuiper, G. P. eds., 1963, *The Moon, Meteorites, and Comets*, Chicago: Univ. of Chicago Press.

Miller, W. C., 1980, 'Dark Adaptation: Its Nature and Preservation', *AAS Photobulletin*, 1980 No. 2 (Consecutive No. 24), p. 18, Working Group on Photographic Materials, AAS, Rochester, Kodak.

Millman, P. M., 1956 August, 'A Complex Perseid Meteor Spectrum', *Sky and Telescope*, p. 445.

Millman, P. M., 1957 May, 'An IGY Program of Meteor Observing for Amateurs', *Sky and Telescope*, p. 317.

Millman, P. M., 1980, 'Meteors, Fireballs and Meteorites', *Observer's Handbook 1981* (updated and issued annually), Percy, J. R., ed., Toronto: Univ. of Toronto.

Millman, P. M. and Clifton, K. S., 1979 January, 'Video Techniques in Comet-Debris Studies', *Sky and Telescope*, p. 21.

Moore, P., ed., 1963, *A Handbook of Practical Amateur Astronomy*, New York: W. W. Norton.

'More About the Earth's Cloud Satellites', *Sky and Telescope*, 1961 August, p. 63.

Morris, C. S., 1979 April 27, 'A New Method for Estimating Cometary Brightness', *Comet News Service*, No. 79-1.

Morris, C. S., 1980 October, 'A Review of Visual Comet Observing Techniques, I', *International Comet Quarterly*, p. 69.

Morris, C. S., 1981a January, 'A Review of Visual Comet Observing Techniques, II', *International Comet Quarterly*, p. 3.

Morris, C. S., 1981b July, 'A Review of Visual Comet Observing Techniques, III', *International Comet Quarterly*, p. 89.

Morris, C. S. and Green, D. W. E., 1982 June, 'The Light Curve of Periodic Comet Halley 1910 II', *Astronomical Journal*, Vol. 87 No. 6, p. 918.

NGC 2000.0, see Sinnott, R. W.

Newburn, R. L., Jr., Bell, J. F., and McCord, T. B., 1981 March, 'Interference Filter Photometry of Periodic Comet Ashbrook–Jackson', *Astronomical Journal*, Vol. 86, No.3, p. 469.

Newburn, R. L., Jr., and Yeomans, D. K., 1982, 'Halley's Comet', in *Ann. Rev. of Earth and Planetary Science*, Vol. 10, Wetherill, G. W., Albee, A. L., Stehli, F. G., eds., Palo Alto: Annual Reviews Inc., p. 297.

'New Natural Satellites of the Earth?' in 'News Notes', *Sky and Telescope*, 1961 July, p. 10.

Nye, R. A., 1981 November, 'Arizona Amateur's Photoelectric Photometer' in 'Gleanings for ATM's', *Sky and Telescope*, p. 496.

Olsson-Steel, D., 1988, 'Identification of Meteoroid Streams From Apollo Asteroids in the Adelaide Radar Orbit Surveys', *Icarus*, Vol. 75, p. 64.

Osborn, W. H., A'Hearn, M. F., Carsenty, U., Millis, R. L., Schleicher, D. G., Birch, P. V., Moreno, H., and Gutierrez-Moreno, A., 1990, 'Standard Stars for Photometry of Comets', *Icarus*, Vol. 88, p. 228.

Owen, M. R., 1986 June, 'VHF Meteor Scatter – An Astronomical Perspective', *QST*, p. 14.

Pansecchi, L., 1981 July, 'Decentering a Lens for Comet Photography', in 'Gleanings For ATMs', *Sky and Telescope*, p. 70.

Patterson, G. N., 1981, *Handbook of Astrophotography for Amateur Astronomers*, Saskatoon: RASC Saskatoon Centre.

Patterson, J. and Michaud, P., 1980 February, 'Photographing Stellar Spectra', in 'Photography in Astronomy', *Astronomy*, p. 39.

Paul, H. E., 1960, *Outer Space Photography for the Amateur*, New York: Amphoto.

Peltier, L. C., 1965, *Starlight Nights: The Adventures of a Star-Gazer*, New York: Harper and Row.

Peltier, L. C., 1972, *Guideposts to the Stars*, New York: Macmillan.

Peltier, L. C., 1977, *The Place on Jennings Creek*, Chicago: Adams Press.

Penhallow, W. S., 1978 November, 'Notes on a 16-inch Astrometric Reflector' in 'Gleanings for ATMs', *Sky and Telescope*, p. 455.

Perrine, C. D., 1934, 'Observaciones del Cometa Halley Durante su Aparicion 1910', *Pub. Resultados del Observatorio Nacional, Argentino*, Vol. 25.

Pilon, K., 1984 May, 'Meteor Astronomy at Home' in 'Computer Adventures', *Popular Science*, p. 80.

Pocock, E., 1992 July, 'The Perseids Meteor Storm', *QST*, p. 29.

Polman, J., 1977 May, 'A Homemade Filar Micrometer' in 'Gleanings for ATMs', *Sky and Telescope*, p. 391.

Povenmire, H. R., 1980, *Fireballs, Meteors, and Meteorites*, Indian Harbor Beach, FL: JSB Enterprises, Inc.

'Projects for May with a Sky Crossbow' in 'Celestial Calendar', *Sky and Telescope*, 1981 May, p. 417.

Rahe, J., Donn, B., and Wurm, K., 1969, *Atlas of Cometary Forms*, Washington, DC: NASA, SP-198.

Rao, J., 1993 August, 'Storm Watch for the Perseids', *Sky and Telescope*, p. 43

Reynolds, M. and Parker, D., 1988 June, 'Hypered Film for Planetary Photography' in 'Observer's Page', *Sky and Telescope*, p. 668.

Richardson, R. S., 1967, *Getting Acquainted with Comets*, New York: McGraw-Hill Book Co.

Ridley, H. B., 1984 December, 'The Photography of Comets', *Journal of the British Astronomical Association*, Vol. 95, No. 1, p. 8.

Roach, F. E. and Jamnick, P. M., 1958 February, 'The Sky and Eye', *Sky and Telescope*, p. 164.

Roach, F. E. and Van Biesbroeck, G., 1954 March, 'The Zodiacal Light and the Solar Corona', *Sky and Telescope*, p. 144.

Roggemans, P., ed., 1987, *Handbook for Visual Meteor Observations*, Cambridge, MA: International Meteor Organization and Sky Publishing Corp.

Romer, J., 1984 September, 'Cheap Astronomy: The Joy of Sketching' in 'Observer's Page', *Sky and Telescope*, p. 277.

Röser, S. and Bastian, U., 1991 and 1993, *Positions and Proper Motion Star Catalogue North* and *Positions and Proper Motion Star Catalogue South*, Heidelberg, Berlin, and New York: Spektrum Akademischer Verlag. The catalogs are also available from the US National Space Science Data Center (NSSDC).

Roth, G. D., ed., 1975, *Astronomy: A Handbook*, Cambridge, MA: Sky Publishing Corp.

Russell, J. A., 1959 August, 'Some Perseid Spectra of the Past Decade', *Sky and Telescope*, p. 549.

Saint Exupéry, A., 1943, *The Little Prince*, New York: Harcourt Brace Jovanovich, p. 14.

Schmadel, L. D., 1993, *Dictionary of Minor Planet Names*, 2nd Edn., New York: Springer Verlag.

Schmidt, T. E., 1984 September, 'Super Quadrantid Spectra' in 'Letters', *Sky and Telescope*, p. 197.

Schmiedeck, W., 1979 April, 'A Simple Technique for Recording the Sun's Spectrum', in 'Gleanings for ATMs', *Sky and Telescope*, p. 395.

Scovil, C. E., 1990, *The AAVSO Variable Star Atlas*, 2nd edn., Cambridge, MA: Sky Publishing Corp., 178 charts.

'Search Ephemeris for the L4 Cloud Satellite' in 'Observer's Page', *Sky and Telescope*, 1962 December, p. 356.

Seargent, D. A., 1982, *Comets: Vagabonds of Space*, Garden City, NY: Doubleday & Co., Inc.

Sekanina, Z., 1993, 'Disintegration Phenomena Expected During the Forthcoming Collision of Periodic Comet Shoemaker–Levy 9 with Jupiter', submitted to *Science*.

Setteducati, A. F., 1960 April, 'Listening to Meteors on Short Wave' in 'Observer's Page', *Sky and Telescope*, p. 363.

Sherrod, P. C., *Observations of Meteors*, Little Rock, Arkansas, USA: The Astronomical Unit.

Sherrod, P. C., *Studies of the Solar System Part I – Comets*, Little Rock, AR: The Astronomical Unit.

Sherrod, P. C. with Koed, T. L., 1981, *A Complete Manual of Amateur Astronomy*, Englewood Cliffs, NY: Prentice-Hall, Inc.

Sidgwick, J. B., 1980, *Amateur Astronomer's Handbook*, 4th edition, Hillside, NJ: Enslow.

Sidgwick, J. B., 1982, *Observational Astronomy for Amateurs*, 4th edition, Hillside, NJ: Enslow.

Simpson, W., 1967 February, 'Dust Cloud Moon of the Earth', *Physics Today*, Vol. 20, No. 2, p. 39.

Sinnott, R. W., ed., 1988, *NGC 2000.0*, Cambridge, MA: Sky Publishing Corp. and Cambridge: Cambridge University Press.

Smithsonian Astrophysical Observatory Star Catalog, Washington, D.C.: Smithsonian Institution, 1966.

Smithsonian Astrophysical Observatory Star Atlas of Reference Stars and Nonstellar Objects, Cambridge, MA: M.I.T. Press, 1969.

Struve, O., 1956 September, 'The Spectra of Comets', *Sky and Telescope*, p. 489.

Struve, O., 1960 February, 'Visual Observations of Meteors', *Sky and Telescope*, p. 200.

Suzuki, K., Magafuji, N. and Kinoshita, M., 1976 May, 'Recording Meteor Echoes By FM Radio' in 'Observer's Page', *Sky and Telescope*, p. 359.

Swenson, G. W., Jr., 1978 May to Oct., 'An Amateur Radio Telescope' series, *Sky and Telescope*; published as a book by Pachart Publishing House, Tucson, AZ.

Swenson, G. W., Jr., 1979 April, 'Antennas for Amateur Radio Interferometers', *Sky and Telescope*, p. 338.

Swenson, G. W., Jr. and Franke, S. J., 1979 November, 'An R. F. Converter for Amateur Radio Astronomy', *Sky and Telescope*, p. 422.

Sykes, M. V. and Walker, R. G., 1992, 'Cometary Dust Trails', *Icarus*, Vol. 95, p. 180.

Tatum, J., 1982a April, 'The Measurement of Comet Positions', *Journal of the Royal Astronomical Society of Canada*, Vol. 76, No. 2, p. 97.

Tatum, J., 1982b June, 'The Calculation of Comet Ephemerides', *Journal of the Royal Astronomical Society of Canada*, Vol. 76, No. 3, p. 157.

Tirion, W., 1981, *B.A.A. Star Charts 1950.0*, London: British Astronomical Association, 5 charts.

Tirion, W., 1981, *Sky Atlas 2000.0*, Cambridge, MA: Sky Publishing Corp.

Tirion, W., Rappaport, B., Lovi, G., 1987, *Uranometria 2000.0*, Richmond, VA: Willmann-Bell, Inc.

Vehrenberg, H., 1963, *Photographic Star Atlas – The Falkau Atlas*, Dusseldorf: Treugesell, 464 charts.

Vehrenberg, H., 1971, *Atlas Stellarum 1950.0*, Dusseldorf: Treugesell, 486 charts.

Waber, R. and McPherson, R., 1967 May, 'Photographing Star Spectra', in 'Observer's Page', *Sky and Telescope*, p. 322.

Watson, F.G., 1962, *Between the Planets*, Garden City, NY: Doubleday and Co., Inc.

Weinberg, J. L., 1967, *The Zodiacal Light and the Interplanetary Medium*, Washington, DC: NASA, SP-150.

Whipple, F. L., 1981, 'On Observing Comets for Nuclear Rotation', in *Modern Observational Techniques for Comets*, Brandt, J. C., Donn, B., Greenberg, J. M., and Rahe, J., eds., Pasadena: NASA-JPL Publication 81–68.

Whipple, F. L., 1985, *The Mystery of Comets*, Washington, DC: Smithsonian Institution Press.

Whipple, F. L. and Hamid, S., 1950, 'On the Origin of the Taurid Meteors', *Astronomical Journal*, Vol. 55, p. 185.

Whipple, G. M., 1884 March 13, 'The Zodiacal Light' (letter), *Nature*, p. 453.

Williamson, I. K., 1966 September, 'Meteor Observations', *Skyward*, the monthly newsletter of the Montreal Centre of the Royal Astronomical Society of Canada.

Williamson, I. K., 1966 October, 'The Perseid Meteor Shower', *Skyward*, the monthly newsletter of the Montreal Centre of the Royal Astronomical Society of Canada.

Wood, F. B., ed., 1963, *Photoelectric Astronomy for Amateurs*, New York: Macmillan Co.

Worley, C. E., 1961 September, 'The Construction of a Filar Micrometer', *Sky and Telescope*, p. 140.

Yeomans, D. K., 1981a, *The Comet Halley Handbook*, Pasadena: NASA – JPL Publication 400–91.

Yeomans, D. K., 1981b, 'Comet Tempel–Tuttle and the Leonid Meteors', *Icarus*, Vol. 47, p. 492.

Yeomans, D. K., 1991, *Comets: A Chronological History of Observation, Science, Myth, and Folklore*, New York: John Wiley & Sons, Inc.

Yeomans, D. K. and Kiang, T., 1981, 'The Long Term Motion of Comet Halley', *Monthly Notices of the Royal Astronomical Society*, Vol. 197, p. 633.

Zahnle, K., Low, M.-M. M., Chyba, C. F., 1993, 'Some Consequences of the Collision of a Comet and Jupiter', submitted to *Nature*.

Index

242